AGRICULTURE ISSUES AND POLICIES

GLYPHOSATE

CHEMISTRY, USES AND SAFETY CONCERNS

Agriculture Issues and Policies

Additional books and e-books in this series can be found on Nova's website under the Series tab.

AGRICULTURE ISSUES AND POLICIES

GLYPHOSATE

CHEMISTRY, USES AND SAFETY CONCERNS

RON QUINN
EDITOR

nova
science publishers
New York

NOTICE TO THE READER

The Publisher has taken reasonable care in the preparation of this book, but makes no expressed or implied warranty of any kind and assumes no responsibility for any errors or omissions. No liability is assumed for incidental or consequential damages in connection with or arising out of information contained in this book. The Publisher shall not be liable for any special, consequential, or exemplary damages resulting, in whole or in part, from the readers' use of, or reliance upon, this material. Any parts of this book based on government reports are so indicated and copyright is claimed for those parts to the extent applicable to compilations of such works.

Independent verification should be sought for any data, advice or recommendations contained in this book. In addition, no responsibility is assumed by the publisher for any injury and/or damage to persons or property arising from any methods, products, instructions, ideas or otherwise contained in this publication.

This publication is designed to provide accurate and authoritative information with regard to the subject matter covered herein. It is sold with the clear understanding that the Publisher is not engaged in rendering legal or any other professional services. If legal or any other expert assistance is required, the services of a competent person should be sought. FROM A DECLARATION OF PARTICIPANTS JOINTLY ADOPTED BY A COMMITTEE OF THE AMERICAN BAR ASSOCIATION AND A COMMITTEE OF PUBLISHERS.

Additional color graphics may be available in the e-book version of this book.

Library of Congress Cataloging-in-Publication Data

ISBN: 978-1-53614-002-6

Published by Nova Science Publishers, Inc. † New York

Contents

PREFACE

Chemical herbicides are widely used in food production throughout the world. At the present time, they represent approximately 40% of chemical products sales, followed by insecticides, fungicides, and other types of pesticides. Among these products, commercial formulations based on N-phosphonomethyl-glycine (PMG), commonly known as glyphosate, are the most used worldwide. Such herbicidal formulations play a key role in promoting crop yields. In *Glyphosate: Chemistry, Uses and Safety Concerns* review information on the presence of glyphosate-based herbicides in the environment, their potentially harmful effects, their influence on soil microbial communities and their capacity to adsorb to clay particles (which affects their environmental availability).

The objective of the following work was to investigate the best way to achieve slow release of glyphosate using layered double hydroxides (LDHs) to minimize the environmental impacts of this herbicide. LDHs are also called anionic clays because they can host negatively charged species between their layers. They are best stabilized by anions with high charge density, such as carbonate. Understanding the release dynamics of glyphosate in homogeneous solutions of anions that are commonly found in the soil is important, since many crops require prior soil correction, altering the pH and also the availability of anions.

In one study, the chelating capacity, coordination modes and structural chemistry of glyphosate with Ni (II) cations were investigated in solution and in the solid state. Glyphosate was purified from commercial Roundup WG (Monsanto), and characterized by 31P NMR, FTIR and melting point. The nickel complexes were prepared from Ni(NO3)2 solutions at pH = 8.0 and 4.0, and characterized by solid-state infrared spectroscopy (4000-150 cm-1) and elemental analysis (CHN). The solids were also studied by thermal analysis and X-ray absorption spectroscopy (XAS) using synchrotron radiation (LNLS, Brazil).

Studies show that glyphosate has an oral absorption of 20 to 40% and is considered to be of low toxicity to mammals; however, recent research has shown that it can induce serious damage to mammalian cells. The authors examine suggested mechanisms of toxicity, including the blocking of mitochondrial oxidative phosphorylation, inhibition of cytochrome p450 activity, inhibition of intestinal arylhydrocarbohydroxylase activity, changes in glucose 6-phosphate dehydrogenase activity, DNA damage, hormonal changes, channel openings of calcium, and neurotoxicity involving NMDA receptor activation.

The widespread and frequent use of glyphosate has increased concern about potential adverse effects on human health. Due to more stringent legislation on wastewater discharge and purity of drinking water, the authors discuss the development of an effective remediation technology for the decontamination of glyphosate. Although glyphosate residues can efficiently be removed by advanced technologies, the treatments are usually expensive and difficult to maintain. Therefore, different methods and options should be considered.

Chapter 1 - Chemical herbicides are widely used throughout the world in food production. At the present time, they represent approximately 40% of chemical products sales, followed by insecticides, fungicides, and other types of pesticides. Among these products, commercial formulations based on N-phosphonomethyl-glycine (PMG), commonly known as glyphosate, are the most used worldwide. Such herbicidal formulations play a key role in promoting crop yields. PMG is a broad-spectrum compound that kills weeds that compete with crops in agricultural soils (such as soils for soybean and maize growth) and non-agricultural environments (such as green spaces and gardens). In the last decade, some annual broadleaf weeds have developed resistance to PMG. For this reason, it has become necessary to apply higher and repeated doses of PMG, which may lead to contamination of the environment. Some works have reported a relatively fast biodegradation of PMG in contaminated soils, with a half-life period of nearly 20 days. However, after 30-years of application of glyphosate-based herbicides (GBH), this herbicide and its primary degradation product, aminomethylphosphonic acid (AMPA), still are the most frequently detected compounds in water bodies, mainly in agricultural areas. In addition to environmental pollution and human health risks, these residual compounds can have a phytotoxic effect on the next crop and become a threat to food production. Several processes act in the dissipation of organophosphorus compounds on soil, being the microbial activity one of the most relevant. Since the 1980s, applied researchers have become increasingly interested in the study of the effects of GBH on microbial communities. Although considerable research has been done on biodegradation of PMG by bacteria, much less is known regarding this capacity in fungi. Fungal species belonging to genera such as *Phanerochaete*, *Trametes*, *Aspergillus*, *Fusarium*, *Penicillium* and *Trichoderma* are able to metabolize important environmental pollutants such as herbicides. They possess a number of advantages that can be exploited in bioremediation systems. They can degrade a diverse range of persistent toxic environmental pollutants and carry out a relevant role in co-metabolic degradation processes. The persistence of PMG and AMPA in soils and aquatic sediments depends on the type of soil, tillage, and climatic conditions, among other factors. Both are polar compounds that strongly bind to Fe and Al oxides, clays and organic matter (humic acids) of the soil, decreasing the availability of free herbicide. The purpose of this chapter is to review information on the presence of GBH in the environment, their potentially harmful effects, their influence on soil microbial communities and their capacity to adsorb to clay particles (which affects their environmental availability).

Chapter 2 - The objective of this work was to investigate the best way to achieve slow release of glyphosate using layered double hydroxides (LDHs), to minimize the environmental impacts of this herbicide. Understanding the release dynamics of glyphosate in homogeneous solutions of anions that are commonly found in the soil is important, since many crops require prior soil correction, altering the pH and also the availability of anions. LDHs are best stabilized by anions with high charge density, such as carbonate. In turn, glyphosate is a potential intercalant due to its charge density. At pH 9, glyphosate is deprotonated both in its carboxylate and phosphonate portions, so it can be intercalated to form a hybrid material with unique properties. The LDH used in this study was prepared by a coprecipitation process with variable pH, and was characterized by infrared spectroscopy, X-ray powder diffraction and ^{27}Al and ^{31}P solid-state nuclear magnetic resonance. The release kinetics of glyphosate was analyzed in solutions containing carbonate, nitrate and bicarbonate anions, either by the colorimetric method of Ruhemann and by HPLC. Both detection methods indicated that the carbonate anions promoted greater release of glyphosate.

Chapter 3 - The Ni (II) cation is essential for symbiotic N_2 fixation by higher plants, and there are conflicting reports in the literature concerning a possible correlation of the continuous use of glyphosate on crops with the decrease of Ni (II) bioavailability in the soil, impairing the growth of some crops, including genetically modified and glyphosate-resistant soybeans. On the other hand, Kutman et al. (2013) reported that foliar nickel application alleviates detrimental drift effects of glyphosate on wheat plants. However, the coordination and structural chemistry of Ni (II) – glyphosate complexes at different pH values and Ni (II): glyphosate stoichiometric ratios have been little studied. In this contribution, the chelating capacity, coordination modes and structural chemistry of glyphosate with Ni (II) cations were investigated in solution and in the solid state. Glyphosate was purified from commercial Roundup WG (Monsanto), and characterized by ^{31}P NMR, FTIR and melting point. The nickel complexes were prepared from $Ni(NO_3)_2$ solutions at pH = 8.0 and 4.0, and characterized by solid-state infrared spectroscopy (4000-150

cm^{-1}) and elemental analysis (CHN). The solids were also studied by thermal analysis and X-ray absorption spectroscopy (XAS) using synchrotron radiation (LNLS, Brazil). The monitoring of the reactions in solution, through the analysis of the ligand field transition bands ($^3A_{2g} \rightarrow {}^3T_{2g}$) by electronic spectroscopy, revealed that complexation occurs in the experimental conditions employed. Infrared spectroscopy showed that in all cases the carboxylate group of glyphosate coordinates in a monodentate fashion, and coordination of the phosphonate and amino groups also occurs in basic media. In some cases, there was indication that water molecules participate in the coordination. It was possible to propose structures from the data for the products of the reactions between glyphosate and Ni (II) at pH = 8.0 and 4.0. At acid pH, the IR results, elemental and thermal analysis, and XAS spectra suggested that the complexes formed are the same regardless of stoichiometry. The results obtained from the simulation of the EXAFS spectra agreed with the vibrational spectroscopy analysis, in the sense that the Ni (II) complex obtained with 1:1 stoichiometry and acid pH was formed by only one coordination sphere composed of six light atoms, presumably oxygens, while the complex obtained at basic pH with 2:1 stoichiometry, on the other hand, was formed by two units, an anionic $[Ni(glyphosate)_2]^{2-}$ and a cationic $[Ni(H_2O)_6]^{2+}$.

Chapter 4 - Glyphosate-based herbicides (GBH) are the major pesticides used worldwide. Their uses have increased exponentially since their introduction on the market. Residue levels in food, water, and human urine has been detected, raising concerns about potential environmental and human health impacts. Research has shown that glyphosate can induce a broad range of biological effects in exposed organisms, particularly via generating oxidative stress. However, no comprehensive investigation has been undertaken of the global molecular mechanisms of toxicity of glyphosate for mammalian species. Studies show that glyphosate has an oral absorption of 20 to 40% and is considered to be of low toxicity to mammals; however, recent research has shown that it can induce serious damage to mammalian cells. Suggested mechanisms of toxicity include blocking of mitochondrial oxidative phosphorylation, inhibition of cytochrome p450 activity, inhibition of intestinal arylhydrocarbohydroxylase activity, changes in glucose 6-phosphate dehydrogenase activity, DNA damage, hormonal changes, channel openings of calcium, and neurotoxicity involving NMDA receptor activation. These cellular alterations are directly related to the long-term development of teratogenic, tumorigenic, and hepatorenal effects. Therefore, this chapter aims to demonstrate the mechanisms of toxicity involving glyphosate in mammals.

Chapter 5 - Glyphosate is a broad-spectrum, non-selective systemic herbicide used for a great variety of crops control. Glyphosate acid and its salts are moderately toxic compounds in the Environmental Protection Agency (EPA) Toxicity Class II, but the widespread and frequent use of Glyphosate has increased concern about potential adverse effects on human health. Due to more stringent legislation on wastewater discharge and purity of drinking water, attention is given to the development of an effective remediation technology for the decontamination of Glyphosate. Although Glyphosate residues can efficiently be removed by advanced technologies, the treatments are usually expensive and difficult to maintain. Therefore, different methods and options should be considered while it should be noted that most likely a single method is not sufficient and the key treatment is by integrated methods. This review discusses the old-to-recent papers on environmental risk, analysis and variety of feasible remediation to remove Glyphosate from water and wastewater.

In: Glyphosate
Editor: Ron Quinn

ISBN: 978-1-53614-002-6

Chapter 1

GLYPHOSATE AND FUNGI IN AGRICULTURAL ENVIRONMENTS

Carina E. Magnoli[1,*], Carla L. Barberis[1], Stella M. Chiacchiera[2], María del Pilar Monge[2], Cecilia S. Carranza[1], Nicolás Benito[1] and Melisa E. Aluffi[1]

[1]Departamento de Microbiología e Inmunología, Facultad de Ciencias Exactas, Físico, Químicas y Naturales, Universidad Nacional de Río Cuarto, Córdoba, Argentina
[2]Departamento de Química, Facultad de Ciencias Exactas, Físico, Químicas y Naturales, Universidad Nacional de Río Cuarto, Córdoba, Argentina

ABSTRACT

Chemical herbicides are widely used throughout the world in food production. At the present time, they represent approximately 40% of chemical products sales, followed by insecticides, fungicides, and other types of pesticides. Among these products, commercial formulations

* Corresponding Author Email: cmagnoli@exa.unrc.edu.ar.

based on N-phosphonomethyl-glycine (PMG), commonly known as glyphosate, are the most used worldwide. Such herbicidal formulations play a key role in promoting crop yields. PMG is a broad-spectrum compound that kills weeds that compete with crops in agricultural soils (such as soils for soybean and maize growth) and non-agricultural environments (such as green spaces and gardens). In the last decade, some annual broadleaf weeds have developed resistance to PMG. For this reason, it has become necessary to apply higher and repeated doses of PMG, which may lead to contamination of the environment. Some works have reported a relatively fast biodegradation of PMG in contaminated soils, with a half-life period of nearly 20 days. However, after 30-years of application of glyphosate-based herbicides (GBH), this herbicide and its primary degradation product, aminomethylphosphonic acid (AMPA), still are the most frequently detected compounds in water bodies, mainly in agricultural areas. In addition to environmental pollution and human health risks, these residual compounds can have a phytotoxic effect on the next crop and become a threat to food production. Several processes act in the dissipation of organophosphorus compounds on soil, being the microbial activity one of the most relevant. Since the 1980s, applied researchers have become increasingly interested in the study of the effects of GBH on microbial communities. Although considerable research has been done on biodegradation of PMG by bacteria, much less is known regarding this capacity in fungi. Fungal species belonging to genera such as *Phanerochaete*, *Trametes*, *Aspergillus*, *Fusarium*, *Penicillium* and *Trichoderma* are able to metabolize important environmental pollutants such as herbicides. They possess a number of advantages that can be exploited in bioremediation systems. They can degrade a diverse range of persistent toxic environmental pollutants and carry out a relevant role in co-metabolic degradation processes. The persistence of PMG and AMPA in soils and aquatic sediments depends on the type of soil, tillage, and climatic conditions, among other factors. Both are polar compounds that strongly bind to Fe and Al oxides, clays and organic matter (humic acids) of the soil, decreasing the availability of free herbicide. The purpose of this chapter is to review information on the presence of GBH in the environment, their potentially harmful effects, their influence on soil microbial communities and their capacity to adsorb to clay particles (which affects their environmental availability).

Introduction

N-phosphonomethyl-glycine (PMG), commonly known as glyphosate, is the active ingredient of several commercial herbicides and is used as a non-selective herbicide to control weeds in agricultural and non-agricultural areas. Despite glyphosate-based herbicides (GBH) are widely used in agriculture, forestry and gardens, no precise data exists on their effects on non-target organisms [1]. PMG is never used alone but in combination with adjuvants such as polyoxiethyleneamine (POEA), which help in the absorption of the active ingredient into plant tissues and cells. These adjuvants can also be toxic. In fact, several toxic effects on humans and the environment have been reported. Commercial formulations of PMG are confidential. Thus, this fact does not allow scientists to describe their mechanism of action neither on non-target organisms on plants. During the last years, public concern has risen on the potential environmental and health problems linked to the agricultural use of PMG, mainly due to the vast areas of application and the volumes involved in the production of genetically modified crops [2].

PMG remains unchanged in soils for varying lengths of time, due to its adsorption to clay particles and organic matter present in soils [3]. This property makes this herbicide very persistent in soils and sediments. The high persistence of PMG and/or aminomethylphosphonic acid (AMPA; its main degradation metabolite) in soils and water surfaces, together with evidences of cancer and endocrine disruption in humans, as well as acute and chronic diseases affecting aquatic species [4-6] lead to a strong discussion concerning the potential toxicity of PMG [7-9].

In this scenario, it becomes necessary to establish specific regulations on the use of GBH and on the promotion of interdisciplinary strategies aimed to enhance the natural processes involved in PMG degradation in soils and water [10].

Glyphosate General Aspects

Nowadays, the intensification of agriculture caused a noticeable dependence on herbicides for weed control [1].

Climate change is among the most important challenges influencing natural ecosystems, agricultural productivity and food security [11].

Climate change not only directly influences on crop plants and forests, but it can also impact on pests of these crops [12]. As the most popular herbicide used worldwide, PMG will probably remain as the world's most used agrochemical during the next years [13]. Despite doubts on the activity of this herbicide in relation to climate change, Jabran and Dogan [14] showed that climatic conditions did not affect its activity. They observed that GBH application provided equal and effective weed control under different CO_2 and temperature levels and their combinations. The sales and use of these herbicides have increased exponentially since 1996, after the introduction of glyphosate-tolerant crop varieties (namely, maize, soybean, canola, cotton, sugar beet, and alfalfa). In the beginning, it was only used in agricultural environments. However, due to their effectiveness, they have been spread into other environments close to urban areas such as green spaces and gardens [2].

Approximately 750 different GBH formulations are marketed worldwide [15]. Indeed, each name of a given formulation represents a different mixture of active ingredient and co-formulations. One of the first and the most known of these herbicides is Roundup, the trade name of Monsanto [16]. These herbicides are mixtures of water with approximately 36-48% of glyphosate as the active ingredient. Like other pesticides, 10-20% of GBH consist on chemical formulations, which are mainly composed by different oxidized petroleum distillates or derivatives such as POEA. These are surfactants, diluents or adjuvants that stabilize PMG and increase the penetration of the active ingredient into plants [17, 18].

These GBH were developed for removing herbs and shrubs, especially perennials that compete with crops. PMG is a non-selective herbicide with systemic action, which is absorbed mainly by the green parts of plant tissues. This compound causes general contamination of the organism, which induces a necrosis of the tissues and leads to plant death. The effectiveness of PMG is associated with the inhibition of the enzyme 5-enolpyruvylshikimate-3-phosphate synthase (EPSPS). Thus, PMG stops the sixth step in the shikimate pathway (conversion from shikimate-3-phosphate to EPSP), which is required for the production of aromatic amino acids and secondary compounds. The application of GBH results in the yellowing and decaying of weeds' leaves within 5–10 days (sometimes 30 d), this being caused by the breakdown of aromatic amino acids synthesis [19, 20].

Application and Persistence of Glyphosate

GBH are intensively applied into agricultural fields before crop planting as well as pre-or post-harvest, both in conventional or in reduced tillage to control the growth of annual and perennial weeds. The large extensions of areas cultivated with genetically modified soybean, canola, cotton and corn tolerant to PMG in Argentina, Brazil and USA, strongly contribute to the high amounts of GBH applied every year worldwide [2]. In Europe, where no genetically modified crops are used, GBH are mainly applied to cereals (wheat, rye, triticale, barley and oats), oilseeds (rapeseed, mustard seed and linseed) and vineyards. In these countries, GBH are usually applied one to three times a year at recommended rates, which are lower than those applied in South American countries [21]. The last are recognized as the main producers of cereals and oilseeds, and their crops are destined mainly for the international market. Together with this production, it should be considered the intensive horticulture involving a wide geographical distribution near urban centers. These extensive cereals and oilseeds productions respond to an agricultural model based on PMG-resistant transgenic crops (mainly soybeans, maize and sunflower), zero tillage, and to an intensive use of fertilizers and other pesticides. This system depends not only on chemical products but also on modern technologies [22].

In this production model, a great expansion of the agricultural frontiers, boosted by the adoption of soybean as main crop, took place in several South American countries such as Brazil, Bolivia, Paraguay and Uruguay. GBH are not only used for weed control, but also as chemical fallow in PMG sensitive crops, such as wheat, among others. In addition, GBH doses increased four-fold between 1996 and 2012, and several cases of weed resistance have been reported [23]. These products have also been applied for fruits and vegetables production considering small-scale activity or family production [24]. Several kinds of vegetables are sowed nearby cities and these areas are known as greenbelts where pesticides are frequently applied. Therefore, the population is increasingly being exposed to these compounds. On the other hand, it is estimated that a significant percentage of pesticides (~ 47%) does not reach the target and remains in adjacent soils, water bodies or in the atmosphere [25].

Currently, the horticultural productivity is also extremely limited by pests and diseases in the subtropical regions of the world. The market priority is the generation of quality products based on good shape, color, and the absence of damage. The use of agrochemicals is the most common method used for controlling pests, diseases, and improves the quality of the products. Some authors [26] have established that the use of GBH causes pollution of watercourses. During the last decade, alarming data regarding annual application of GBH have been reported: 281 million liters [27]. Despite the intense use, little is known on the environmental fate of PMG. It has been described that the final destination depends on climatic conditions such as rainfall and temperature, geological features, PMG formulations and presentations and application techniques, among other factors. This herbicide is a polar, highly water soluble substance (11.6 g/L, 25°C) [28], which explains its presence in aquatic systems, mainly around fumigated areas [29]. Despite its persistence is relatively short compared to other pesticides (it has a half-life that ranges from 2 to 91 days), PMG persists more time and it may last up to 215 days when it combines with soil particles or sediments. Similarly, AMPA is more persistent than PMG, with a higher half-life ranging from 76 to 240 days in soil [30]. A frequent occurrence of PMG and AMPA in water surfaces would indicate a continuous application of this herbicide [31].

TOXICITY

Despite the intense use of GBH, the information on their effects on non-target organisms remains incomplete. Side effects are variable since the herbicidal formulations differ among the different trademarks, so the adjuvants have different levels of toxicity alone or in combination with PMG. Therefore, there is an evident problem of reproducibility and consistency in toxicological studies. This could by partly due to the fact that comparisons are performed between different formulations [32]. The information available on PMG toxicity is very controversial, e.g., some researchers established that the use of this herbicide implies no risk for aquatic biota and humans [19, 33, 34], while other authors reported that PMG represents a significant risk to aquatic microorganisms, invertebrates and amphibians, at certain concentrations [26, 28].

The hypothesis of higher toxicity of commercial formulations over PMG alone has been proven in laboratory animals [35], other animal model systems such as sea urchins [36], microorganisms and crustaceans [37]. Mottier et al. [38, 39] showed that POEA-based adjuvants can be very toxic for the embryonic and larval development in two life stages of the Pacific oyster. Metamorphosis tests revealed that the EC50 values for PMG and AMPA exceeded 100,000 µg/L while for some commercial formulations was as low as ~ 6,000 µg/L. The composition and toxicity of the different ingredients of GBH have been studied. Mesnage et al. [18] evaluated the toxicity of different brands of GBH in tissue culture cell assays and showed that several commercial formulations were up to 1,000 times more toxic than the active ingredient "glyphosate."

In a recent study, Defarge et al. [16] evaluated the toxicity of commercial GBH formulations and some of their ingredients alone on plant and human cells. These authors showed that PMG was only slightly toxic on plants at the recommended doses used in agriculture, which contrasted the general belief. The strong herbicidal and toxic properties were exerted only by the compounds belonging to the POEA family. The toxic and endocrine disrupting effects of the formulations were mostly due to the entire formulation and not to PMG. In addition, these authors established the importance of heavy metals such as arsenic, chromium, cobalt, lead and nickel, which are known to be toxic and endocrine disruptors. These contaminants are present in almost 22 pesticides, including GBH. This information has also been documented in the world with other herbicides [40] and emphasizes that the toxic effect of many pesticides is perhaps due to other ingredients and not to the active ingredient itself. GBH formulations at the recommended 1% dilution are more toxic on plants than PMG alone in the short term. This fact has also been previously demonstrated in different human cells models [18, 41, 42]. Defarge et al. [16] also showed that there is no instance in which PMG reaches the toxicity of any ingredient, either in the formulations or alone. The effects of the ingredients were membrane disruption, apoptosis, mitochondrial respiration inhibition [18, 42-45], and DNA damaging [46, 47]. PMG alone appears to enter the cells [48], but its toxic action is not well documented at this level. This toxicity is considered secondary in contrast with the effects observed with the other ingredients. The petroleum oxidative derivatives also penetrate and accumulate due to their chemical amphiphilic nature.

In other studies, Mesnage et al. [49-50] showed that the chronic administration of a GBH induced liver toxic effects in rats at a concentration applied to the environment while informing that a daily intake of the active ingredient was declared safe by regulatory agencies worldwide. However, it remains unclear if PMG, the adjuvants, or their combination are responsible for the liver toxicity observed in the studied animal models. Other authors [51] evaluated the toxicity of four commercial GBH on soil invertebrates. They also established that the toxicity of PMG varied according to the commercial formulation.

Regulation agencies such as the United States Environmental Protection Agency [13] classify glyphosate formulations as scarcely or non-toxic to birds and mammals and non-toxic to moderately toxic to aquatic invertebrates. On the other hand, the World Health Organization classifies this herbicide into group 2A, i.e., probably carcinogenic to humans [8]. According to their detailed reports, this is probably due to the different toxicity profiles of the full formulations and PMG alone. The European Commission recommends to ban the use of POEA-type adjuvants in GBH products. Although banning can be seen as a positive step forward for public health, this does not exclude the use of POEA in other non-GP-based products. POEA is also still authorized as a co-formulation in pesticide formulations containing other ingredients such as 2,4-D. Thus, farmers and the population can still be exposed to POEA despite its ban in GBH. Furthermore, the finding that POEA is widely applied within pesticides in agricultural fields raises concerns that this and other classes of pesticide adjuvants may be entering the food and feed chain with unknown health consequences [32].

In addition, the European Commission based on the European Chemical Agency (ECHA) [52] established that PMG is not proven to be carcinogenic, mutagenic or to negatively affect reproduction. However, it can cause serious eye damage and be toxic to aquatic biota, with long-lasting effects. ECHA's opinion is only based on the evaluation of PMG hazardous properties, without considering its levels in the environment (e.g., atmosphere, aquatic and terrestrial ecosystems) or the exposure and associated risks for humans and wildlife.

Environmental Levels

The environmental fates of PMG and AMPA have been well studied in North America since the 80's [53-55], but the issue has been more recently addressed in South America. In Argentina, environmental concentrations were assessed for the first time in a main agricultural area, less than ten years ago [56]. Recently, concentrations of PMG and AMPA have been reported in surface water and sediments in areas with intensive and extensive agricultural models [57-62]. In addition, Lupi et al. [63] reported a high adsorption of PMG and AMPA on sediments from an agricultural watershed from the Southeastern region of Argentina. These works showed common distribution patterns for both metabolites. Evident differences in the concentrations were found between core and marginal productive areas, indicating that important variations exist among regions. Recently, Bonansea et al. [26] evaluated the presence and distribution of these compounds in fresh water, sediment, and suspended particles in samples from a river running across a large city and through areas with intensive and extensive agriculture. From the total samples, up to 35% contained

PMG, AMPA, or both compounds. Analytes concentrations were found at different percentages depending on the environmental matrices considered. The levels ranged from 12 to 20 times higher for PMG and AMPA in sediment and suspended particles, as compared to the levels found in water. The most polluted area was situated within a greenbelt zone of the city, while in second place was an area with extensive agriculture. These researches suggest a relationship between the occurrence and concentrations of these compounds and the use of the herbicide under current agricultural practices. In addition, higher occurrence and concentrations of PMG and AMPA were found in solid matrices. The high sorption affinity of PMG by the upper soil conditions its retention [64]. However, other studies have also showed the flow of the herbicide through macropores, which would allow the mobilization of the herbicide across the soil particles, and could also explain its occurrence in groundwater from areas located near agricultural basins [65, 66].

In a study carried out by the U.S Geological Survey from 2002 to 2006 [54] with ground and surface-water, rainfall and soil samples, AMPA was detected more frequently and at similar or higher concentrations than the parent compound, PMG. Both compounds were more frequently found in surface water than in ground water and trace levels persisted in the soil from year to year.

In Europe, where genetically modified crops are not allowed, PMG has been detected in several water sources (but at lower levels than in the USA). Recently, a large-scale assessment of PMG and AMPA distribution was carried out in the agricultural topsoils of the European Union (EU). PMG and/or AMPA were present in 45% of the topsoils collected from eleven countries and six crop systems, with a maximum concentration of 2 mg/kg. Several PMG and AMPA hotspots were identified across the EU (obtained from recently derived European maps) and they were used to estimate the potential export of PMG and AMPA by wind and water erosion. The estimated potential exports clearly indicated that particulate transport contributes to human and environmental exposure to herbicide residues [21].

Very low concentrations of PMG (<0.1 to 2.5 µg/L) were detected in samples of surface water in Germany [67], Switzerland [62], Hungary [68] and in groundwater samples from northeastern Spain [69]. Higher levels (up to 165 µg/L) were sporadically found in France [70] and Denmark [71]. Moreover, there is scarce information on PMG residues in the environment in other continents. In particular, hardly any information is publicly available on the environmental residues in China, despite being one of the countries with the highest agricultural production [72].

In food and feed agricultural regulations [73, 74], both PMG and AMPA are considered in residue analyses. The concentrations of PMG plus AMPA vary widely, ranging from 0.1 to 100 mg/kg in legumes (including soybeans), 0.1 to 25 mg/kg in cereals and rice, 0.1 to 28 mg/kg in oil seeds and 1 to 344 mg/kg in several types of fodder [75-78].

The tolerable residue levels in seeds and fodder have increased over time due to the increasing concentrations detected in some farm products [2]. Residues of PMG and AMPA in water and plant products are taken up by animals and humans and excreted [79, 80]. PMG was detected in urine of farm animals and humans in high levels (30–80%) [79, 81, 82]. Residues were found not only in the urine of farmers but also in 60–80% of the general public, including children, in the USA and in 44% of the general population in Europe [79, 81]. Daily exposures to PMG are generally lower than the tolerable reference dose currently demanded by regulatory agencies [80, 83]. However, these tolerance levels and safety standards need modifications in view of the expanded human exposures [84].

Effect of Glyphosate on Microbiota in Agricultural Environments

A significant fraction of pesticides sprayed on crops may be returned to soils via plant residues, but its fate has not been thoroughly documented. Mamy et al. [85] studied the fate of PMG associated to plants residues.

Oilseed rape, considered as a crucifer weed, was used as model plant using two lines, glyphosate-tolerant (GT) and a non-GT line. They found that the mineralization of the herbicide in soil was slower when incorporated into plant residues and that PMG availability for mineralization increased when the size of plant residues decreased. After 80 days of soil incubation, extractable 14C-residues mostly involved one metabolite of PMG (AMPA). However, up to 2.6% of initial 14C was still extracted from un-decayed leaves as PMG. Thus, the trapping of herbicides in plant materials provided a protection against degradation, which may increase the persistence of PMG in soils. This pattern appeared to be more pronounced for GT crops, which accumulated a higher amount of non-degraded PMG into their tissues.

The herbicides can harm non-target organisms [86], which may reduce biodiversity and the provision of ecosystem services that support food security and on-farm profitability. Soil microorganisms play central roles in the degradation of herbicides and drive terrestrial ecosystem functioning [87]. Currently, little information is available concerning the level to which herbicides influence on the diversity and function of fungal populations. The effects of cotton-maize rotation and GBH use on counts of soil-borne *Aspergillus flavus* and levels of aflatoxin and fumonisin contamination in maize and cotton seed were determined by Reddy et al. [88] during 2002-2005 in Stoneville, Mississippi (USA). They studied four rotation systems (continuous cotton, continuous maize, cotton-aize and maize-cotton) for both GT and non-GT cultivars. *A. flavus* populations in surface soil, sampled before planting, ranged from 1.47 to 3.00 $\log_{10}$ CFU/g of soil in the four rotation systems. Propagules of *A. flavus* were higher in the continuous maize system compared to the continuous cotton system, and decreased when cotton was rotated with maize. Propagules of *A. flavus* were significantly greater in plots with GT cultivars compared to non-GT cultivars. In cotton seeds, aflatoxin and fumonisin levels were similar independently of rotation and GBH application. In maize grains, aflatoxin was above the regulatory level only in the GT cultivar, during 2004 and 2005. Fumonisin was higher in non-GT cultivar independently of rotation in 2004. Nevertheless, in 2002, 2003 and 2005, the levels of aflatoxin and fumonisin were similar independently of rotation and GBH application. These results indicate the potential for increased aflatoxin and fumonisin levels in maize. Notably, the climatic conditions recorded during this study were not favourable for mycotoxin production. In laboratory studies, these authors showed that high concentrations of PMG were required for inhibiting the growth of *A. flavus*. However, no short-term effects of soil treatment with PMG (analytical quality) were observed on *A. flavus* populations. These data suggest that altered populations of *A. flavus* or higher aflatoxin concentrations in maize grains were due to indirect effects of the GR cropping system. Similarly, Krzysko-Lupicka and Sudol [89] showed that 1 mM of GBH applied in soils with high carbon content caused a significant change in fungal counts. On the other hand, differences in fungal counts were observed after 180 days of growth in soils with low carbon content. The evaluation of the fungicidal activity of the tested concentrations of GBH showed that the dominating group of fungi belonged to the genus *Fusarium* (*F. solani* and *F. oxysporum*) and found that none of the species were sensitive. These strains showed resistance and ability to degrade this herbicide.

Ratcliff et al. [90] examined changes in the structure of bacterial and fungal communities and utilization of carbon by epifluorescent microscopy and traditional culturing techniques. These authors determined whether application of GBH results in short-term changes, either deleterious or beneficial, in forest soil microbial communities. The application at recommended field rates to a clay loam and a sandy loam forest soil resulted in few changes in the structure of microbial communities. Total and culturable bacteria, fungal hyphal length, bacterial fungal biomass, carbon utilization profiles, and bacterial and fungal phospholipid fatty acids were unaffected at 1, 3, 7, or 30 days after application of a commercial formulation. In contrast, a high concentration of GBH simulating an undiluted chemical spill substantially altered the bacterial community in both soils. An increase in total bacteria and culturable bacteria was shown. Culturable bacteria increased from about 1% of the total population in untreated soil. Community composition in both soils shifted from fungal dominance to an equal ratio of bacteria to fungi.

Functional diversity of culturable bacteria, estimated by carbon substrate utilization, also increased at the highest PMG concentration, particularly in the clay loam soil. Changes in fungal properties (hyphae, propagules, biomarkers) were few and transient. The authors concluded that the commercial formulation of PMG had a benign effect on the communities' structure when applied at the recommended field rate, and produced a non-specific, short-term stimulation of bacteria at a high concentration.

In other study, Zabaloy et al. [91] investigated the impacts of GBH on soil microbial communities by an integrated approach, using short-term soil incubations. The key structural and functional parameters were estimated by culturable aerobic heterotrophic bacterial (AHB) density, substrate-induced respiration (SIR), dehydrogenase activity (DHA), fluorescein diacetate (FDA) hydrolysis, and functional richness. Three different soil types were collected from agricultural fields with reported history of herbicide application. Soil microcosms were treated with one herbicide at a time at a dose 10 times higher than the recommended field application rates (150 mg/kg) and incubated for up to 3 weeks. Several short-term effects of GBH on microbial activities and bacterial densities were observed such as early stimulation of SIR and AHB, dissimilar responses in the soils for FDA and DHA and a transient increase in functional richness. These authors concluded that the addition of these herbicides at a dose 10 times higher than the normal field application rates caused minor changes to soil microbial activity, bacterial density and functional richness. The specific changes varied among herbicides, with the effects of PMG being the most pronounced.

Mijangos et al. [92] studied the short-term effects of GBH on rhizosphere soil microbial communities through the utilization of cultivation-dependent and independent techniques that included two different forage plant species and two concentrations of PMG (50 and 500 mg/kg of soil, as commercial formulations). The results showed that PMG stimulated the rhizosphere soil microbial communities after 15 days of application. The activity and functional diversity of the cultivable portion of the heterotrophic soil microbial community is not surprising. These authors concluded that the PMG might act as an available source of carbon, nitrogen and phosphorous after its rapid adsorption to organic and mineral soil surfaces [93, 94]. Moreover, higher values of soil ammonium content were observed at higher PMG concentrations. In other study, Kremer et al. [95] reported that PMG, when applied to soybean leaves, is translocated through the plant, accumulates primarily in meristematic tissues, and is released into soil from roots, and increases carbohydrate and amino acid contents in root exudates. Similarly, Liu et al. [96] found that PMG had no effect on exudation of aromatic amino acids but tended to increase the total amount of amino acids exuded from plants are able to use PMG as a source of energy and nutrients, whereas it may be toxic to other groups [97]. Toxic effects of PMG on soil microorganisms have been explained by the presence of the EPSPS enzyme that can be found both in plants and soil microorganisms [98, 99].

In microcosms, Imparato et al. [100] studied the indirect effect of GBH on the microbial community in the rhizosphere of barley with application of GBH in the phyllosphere, as compared to leaving the plant intact or cutting the shoot off. Foliar application of the herbicide doubled microbial abundance of the culturable fraction in the rhizosphere compared to the other treatments, which had no effect on total abundance. Overall, their results showed that when applying PMG to barley leaves, the availability of organic carbon in the rhizosphere of the dying roots is altered, which in turn may alter the communities and their interactions. This could have implications for general soil carbon turnover processes and CO_2 release in arable systems.

A field study evaluated the impact of annual GBH applications (sub-lethal and recommended field doses) on arbuscular mycorrhizal fungi, dark septate endophytes and free living diazotrophs in temperate grassland. Arbuscular mycorrhizal fungi viable spores and free-living diazotrophs densities were reduced by 56 and 82% respectively, after the fourth application, even at sub-lethal dose. In addition, total arbuscular mycorrhizal fungi root colonization in *Lolium arundinaceum* was not affected among treatments. However, the arbuscules percentage was reduced in plants grown in plots treated with 3 L/ha of GBH. A similar response was detected in dark septate endophytes root colonization.

Considering the role they have in structuring plant communities, these deleterious effects on beneficial soil microorganisms might negatively impact on grassland productivity and diversity according to Druille et al. [101]. Similarly, Newman et al. [102] investigated shifts in the rhizosphere communities following long-term GBH application on GTcorn and soybean, under greenhouse conditions. In this study, rhizosphere soil was sampled and the composition of microbial communities was compared between PMG treated and untreated rhizospheres. The relative abundances of communities increased in maize and soybean following exposure to PMG. Recently, Denis et al. [103] investigated the effects of one application of GBHs on soil microbial diversity and function. The herbicide was added to the soil at recommended doses, which was incubated under laboratory conditions for 60 days. High through put phylogenetic marker gene sequencing revealed that none of the herbicides significantly influenced the richness, evenness and composition of microbial communities. From a functional perspective, herbicides did not significantly affect hydrolysis of FDA and beta-glucosidase activity. Furthermore, the ability of the soil microorganisms to use 15 substrates was generally unaffected by herbicide application. The only exception to this was a temporary impairment in the ability of soil organisms to use three organic acids and an amino acid.

Aquatic habitats support a specialized community of organisms and are often embedded within and around agricultural lands that are frequently exposed to pesticides. Muturi et al. [104] used a microcosm approach to examine the single and combined effects of two herbicides (atrazine and PMG) on microbial communities of aquatic habitats. They showed that individual herbicides reduced the microbial diversity. A mixture of herbicides had no effect on microbial diversity or structure while a mixture of insecticides or all five pesticides assayed reduced microbial diversity and altered the community structure. This information suggests that the exposure of aquatic ecosystems to individual pesticides or their mixtures can disrupt aquatic microbial communities.

GLYPHOSATE ADSORPTION

Some amount of PMG reaches the soil, where a part is adsorbed for its constituents while other part remains in solution. The biological activity, physicochemical properties of the medium and several chemical reactions will participate in the mineralization of the pesticide [105]. The fate of a pesticide in the soil is governed by the processes of retention, transport, and degradation, as well as by the interaction of these processes. Adsorption is due to the attraction or repulsion between a solid surface and, in this case, a vapor or a solution. This attraction or repulsion results from the interaction between the surface of the adsorbent and the adsorbate´s molecules or ions [106]. The adsorption processes may be either reversible or irreversible and they can retard or prevent the pesticide´s movement, affecting its bioavailability for plants or microorganisms [107].

The mobility, and therefore the leachability, of a compound in the soil depends on the strength of its adsorption to the soil solids that immobilize it, so when more weakly adsorbed it will be leached more easily. Although Rueppel et al. [108] considered PMG as a pesticide with no propensity for leaching, traces of PMG and its major degradation product AMPA were found in surface-water and groundwater environments [56, 109, 110]. The sources and transport routes of pesticides to groundwater are complex. In fact, Landry et al. [109], reported groundwater contamination in shallow soils on the karstic terrain of Vosne-Romanée. On the other hand, Degenhardt et al. [110] found that PMG degradation to AMPA and herbicide sorption to the bottom sediment were more important routes for PMG dissipation than its movement with infiltrating water. In the experiment, ephemeral (E) and semi-permanent (SP) wetlands were divided into halves using a polyvinyl curtain. Half of each wetland was fortified with PMG simulating an accidental direct overspray. PMG dissipated rapidly in the water column of the two prairie wetlands studied. Peruzzo et al. [56] reported temporal variation of the PMG levels with the time of application and the rain events. Thus, levels of PMG were determined in water, soil and sediment samples from a transgenic soybean growing area located near tributaries of the Pergamino-Arrecifes system in the north of the Province of Buenos Aires, Argentina. The fieldwork took into account both pesticide application and the rains that occurred after the applications.

The soil is formed by minerals, organic matter, water, air and living organisms. It has been suggested that soil is the biggest natural bioreactor in the world since it is able to transform, filter, buffer and purify the substances that reach its surface, thus regulating biogeochemical cycles [111].

Compared to less polar pesticides, PMG has unique adsorption characteristics in soils. This herbicide is moderately to weakly adsorbed in soils, mainly organic, since most of their chemical structures are aliphatic and/or aromatic chains with little polar moieties [112, 113]. PMG is a small molecule that contains three polar functional groups (carboxyl, amino and phosphonate). Actually, PMG is a polyprotic acid that forms mono and divalent anions in the range of soils pH (4-8). These anions have strong affinity for trivalent cations such as Al^{3+} and Fe^{3+} [114, 115]. This characteristic has been claimed, at least partially, to be responsible of the strong PMG binding to soil minerals [114].

The mineral fraction of soils is composed of crystalline clay minerals and crystalline and amorphous oxides and hydroxides. Clay minerals such as montmorillonite and vermiculite have a high cation exchange capacity and a high surface area. On the other hand, illite, kaolinite, and chlorite, have lower cation exchange capacities and surface areas. These clays are phyllosilicates with laminar structures. Sepiolite and attapulgite are also phyllosilicates, although they have pseudo-laminar or tubular structures. Smectites is a group of highly expandable clay mineral. Montmorillonite, also known as bentonite, belongs to the smetite group and has a three-layer arrangement, two tetrahedral silicon layers joined together by an octahedral aluminum layer. Although montmorillonite is one of the most predominant clays in some soils, it has been also used as a soil additive to hold soil water in drought-prone soils. Sepiolite is a fibrous clay that occurs mostly in agricultural soils of arid regions. Sepiolite and talc are three layered hydrated magnesium silicates with an inner octahedral magnesium layer and two tetrahedral silicon layers. Attapulgite is found in some calcareous soils and is frequently used as soil improver in potting soils. It is also a three layered clay, with two silicon tetrahedral layers and an aluminum and magnesium inner octahedral sheet. Kaolinite, which occurs in soils formed from the chemical weathering of rocks in hot, moist climates, has only two layers, one octahedral of aluminum and a second, tetrahedral, of silicon. The structural conformation of each clay depends on its chemical composition and also on the configuration under which the minerals are organized within each layer [116].

The binding of organic molecules to clay minerals depends on: a) the characteristics of the silicate (nature, charge density, predominant cation saturating the interlaminar space, among others); b) the temperature of the whole system and c) the properties of the adsorbed molecules (functional groups, size, protonation capacity, among others).

PMG is moderately adsorbed by silicate clays and the level of adsorption seems to depend on the amount of cations in solution. The adsorption of PMG on montmorillonite is particularly interesting since this clay provides different sites available for adsorption [117]. In order to understand the expected site for PMG adsorption, the net clay surface charges have to be considered. Two different types of charges can be distinguished: (i) permanent negative charges in the basal plane (i.e., the "faces") due to the isomorphic substitution of Si (which belongs to group 4A) by Al (group 3A) in tetrahedral sheets of SiO_4; and (ii) the pH-dependent loading on the "edges" due to the reversible protonation-deprotonation of the hydroxyl groups [118]. The negative pH independent charge, due to the morphology of the plate, represents more than 90% of the montmorillonite surface charge [119]. At low pH (i.e., high proton activity), the edge charge becomes less negative and even positive according to the reversible protonation and deprotonation reactions of surface silanol (Si-OH) and aluminol (Al-OH) groups. Therefore, isomorphic substitution in the monmorillonite structure creates negative binding sites over a wide range of pH while surface hydroxyl groups are responsible for the formation of surface complexes with metal ions.

The anionic structure of PMG cannot be adsorbed onto permanent-negative charge sites on the montmorillonite surface at soil pH. Abend and Lagaly [117], suggested that the main mechanism of adsorption would be by ligand exchange and would be favored at acidic pH. However, the coordination with polyvalent cations (A^{3+}, Fe^{3+}) adsorbed to the basal planes of minerals cannot be disregarded [120, 121]. The pH of the medium would favor one or another interaction route [122]. Indeed, adsorption studies with montmorillonite, illite and kaolinite, showed an increased PMG sorption on saturation with divalent and trivalent cations [123]. A complex formation between interlayer cations and PMG was assumed to be the responsible for the adsorption to montmorillonite [124]. In fact, PMG forms stable tridentate complexes in aqueous solution [125]. Although the interlayer complex formation is a probable explanation for adsorption by montorillonite, it does not allow explaining the observed adsorption by illite, with no available interlayer, or by Kaolinite, which lacks an interlayer [126]. On the other hand, Morillo et al. [127] reported that the presence of Cu^{2+} decreases PMG adsorption by montmorillonite, and proposed Al-OH end groups as the potential binding sites. The effect of Cu^{2+} was explained by the formation of Cu-lyphosate complexes with lower surface affinity than PMG. Borggaard and Gimsing [128] suggested that the OH groups on octahedral layers, similar to the singly coordinated hydroxyl groups on aluminium and iron oxides, contribute to PMG sorption by clay silicates. In fact, soils enriched with the variable-charge minerals (allophane/imogolite and goethite) have been proven to be effective PMG sorbents, whereas soils dominated by permanent charge minerals such as illite, smectite and vermiculite adsorb lower amounts of PMG.

The adsorption of PMG by organic matter and competition by soil phosphates should be considered. Soil organic matter (SOM) plays a controversial role in the adsorption of PMG. On one hand, studies have shown that it is negatively correlated with the SOM content [129, 130]. However, Piccolo et al. [131], observed a very high adsorption in four samples of humus. Studies on the competition between glyphosate and phosphate for soil sorption sites concluded that the phosphate content of soils was the most decisive factor for glyphosate sorption. In soils rich in phosphates, this competition can lead to an increased risk of PMG leaching [121, 132, 133].

The diversity of edaphic properties as well as their combination affects the environmental fate of pesticides. The adsorption and leaching of PMG seems to depend on soil characteristics (mineralogical composition, pH, phosphate content and organic matter).

CONCLUSION

The increase in the areas destined to agriculture under the new productive model leads to the use of chemical products, being the herbicides, without doubt, the most used. During the last years, GBH have been extensively studied. Despite their high dissipation rate in the environment, high levels of both PMG and AMPA are detected in countries where the use can be classified as intensive. This indicates that the levels of GBH incorporated into the environment are elevated and exceed the rates of natural degradation. The overview of the presented data indicates the need to extend the studies on the toxic effects and environmental levels of the adjuvants and/or formulated ingredients. In addition, more information on the persistence and toxic effects of AMPA is needed. Currently, there is abundant information on the acute effects of GBH on several organisms. Nevertheless, there is scarce data on the chronic effects of prolonged exposure to low levels of formulated products containing PMG. Similarly, it is also necessary that regulatory agencies control the application of current regulations regarding the use of pesticides in general as well as the levels in agricultural products that reach the consumers. In this context, it is also necessary to develop new strategies that enhance the natural dissipation of GBH. The fungal populations, due to physiological characteristics and metabolic adaptation capacity play a main role in herbicide catabolism and final mineralization. The role of fungi in microbial consortia as well as the effect of environmental factors on the *in situ* degradation also needs further research in order to improve the bioremediation efficiency in natural environments.

REFERENCES

[1] Allegrini, M., Zabaloy, M. C., Gomez, E. and Del, V. (2015). Ecotoxicological assessment of soil microbial communities' tolerance to glyphosate. *Science of the Total Environment*, 533: 60-68.

[2] Benbrook, C. M. (2016). Trends in glyphosate herbicide use in the United States and globally. *Environmental Sciences Europe* 28 (1):3.

[3] Cassigneul, A., Benoit, P., Bergheaud, V., Dumeny, V., Etiévant, V., Goubard, Y., Maylin, A., Justes, E. and Alletto, L. (2016). Fate of glyphosate and degradates in cover crop residues and underlying soil: a laboratory study. *Science of the Total Environment* 545–546: 582–590.

[4] Gasnier, C., Dumont, C., Benachour, N., Clair, E., Chagnon, M. C. and Séralini, G. E. (2009). Glyphosate-based herbicides are toxic and endocrine disruptors in human cell lines. *Toxicology* 262:184-191.

[5] Guyton, K. Z., Loomis, D., Grosse, Y., El Ghissassi, F., Benbrahim-Tallaa, L., Guha, N., Scoccianti, C., Mattock, H. and Straif, K. (2015). Carcinogenicity of tetrachlorvinphos, parathion, malathion, diazinon, and glyphosate. *The Lancet Oncology* 16: 490-491.

[6] Mesnage, R., Arno, M., Costanzo, M., Malatesta, M., Seralini, G. E. and Antoniou, M. N. (2015) Transcriptome profile analysis reflects rat liver and kidney damage following chronic ultra-low dose Roundup exposure. *Environmental Health*, 14: 70.

[7] European Food Safety Authority (EFSA). (2015). Conclusion on the peer review of the pesticide risk assessment of the active substance glyphosate. *EFSA Journal*, 13 (11): 4302 107 pp.

[8] International Agency for Research on Cancer (IARC). (2015). IARC Monographs Volume 112: *Evaluation of five organophosphate insecticides and herbicides.* Available online: http://www.iarc.fr/en/

media-centre/iarcnews/pdf/MonographVolume112.pdf. Access date: 12 March 2018.

[9] Myers, J. P., Antoniou, M. N., Blumberg, B., Carroll, L., Colborn, T., Everett, L. G., Hansen, M., Landrigan, P. J., Lanphear, B. P., Mesnage, R., Vandenberg, L. N., vom Saal, F. S., Welshons, W. V. and Benbrook, C.M. (2016). Concerns over use of glyphosate-based herbicides and risks associated with exposures: a consensus statement. *Environmental Health*, 15: 19.

[10] Van Bruggen, A. H. C., He, M. M., Shin, K., Mai, V., Jeong, K. C., Finckh, M. R. and Morris, J. G. Jr. (2018). Environmental and health effects of the herbicide glyphosate. *Science of the Total Environment* 616–617: 255–268.

[11] Howden, S. M., Soussana, J. F., Tubiello, F. N., Chhetri, N., Dunlop, M., Meinke, H. (2007). Adapting agriculture to climate change. *Proceedings of the National Academy of Sciences,* 104: 19691-19696.

[12] Olesen, J. E. and Bindi, M. (2002). Consequences of climate change for European agricultural productivity, land use and policy. *European Journal of Agronomy*, 16: 239-262.

[13] USEPA (United States Environmental Protection Agency). (2015). *Reregistration Eligibility Decision (Red): Glyphosate*. Office of Prevention, Pesticides and Toxic Substances. Available online: https://archive.epa.gov/pesticides/reregistration/web/pdf/0178fact.pdf (accessed on 26 March 2018).

[14] Jabran, K. and Dogan, M. N. (2018). High carbon dioxide concentration and elevated temperature impact the growth of weeds but do not change the efficacy of glyphosate. *Pest Management Science*, 74: 766–771.

[15] Guyton, K. Z., Loomis, D., Grosse, Y., El Ghissassi, F., Benbrahim-Tallaa, L., Guha, N., Scoccianti, C., Mattock, H. and Straif, K. (2015). Carcinogenicity of tetrachlorvinphos, parathion, malathion diazinon, and glyphosate. *The Lancet Oncology*, 16 (5):490–491.

[16] Defarge, N., Spiroux de Vendômois, J. and Séralini, G. E. (2018). Toxicity of formulants and heavy metals in glyphosate-based herbicides and other pesticides. *Toxicology Reports*, 5: 156–163.

[17] Cox, C. and Surgan, M. (2006). Unidentified inert ingredients in pesticides: implications for human and environmental health. *Environmental Health Perspective*, 114: 1803–1806.

[18] Mesnage, R., Bernay, B., Seralini and G. E. (2013). Ethoxylated adjuvants of glyphosate-based herbicides are active principles of human cell toxicity. *Toxicology*, 313: 122–128.

[19] Giesy, J. P., Dobson, S. and Solomon, K. R. (2000). Ecotoxicological risk assessment for Roundup® herbicide. *Reviews of Environmental Contamination and Toxicology*, 167: 35-120.

[20] Lane, M., Lorenz, N., Saxena, J., Ramsier, C. and Dick, R. P. (2012). The effect of glyphosate on soil microbial activity, microbial community structure, and soil potassium. *Pedobiologia*, 55: 335-342.

[21] Silva, V., Montanarella, L., Jones, A., Fernández-Ugalde, O., Mol, H. G. J., Ritsema, C. J. and Geissen, V. (2018). Distribution of glyphosate and aminomethylphosphonic acid (AMPA) in agricultural topsoils of the European Union. *Science of the Total Environment,* 621:1352-1359.

[22] Lantieri, M. J., Butinof, M., Fernández, R., Stimolo, M. I., Blanco, M. and Díaz, M. P. (2011). Work practices, exposure assessment and geographical analysis of pesticide applicators in Argentina. In: *Pesticides in the modern world-effects of pesticides exposure*. Stoytcheva, M., Editor. InTech: Rijeka, Croatia, pp. 115–138.

[23] Fischer, R. A., Byerlee, D. and Edmeades, G. O. (2014*). Crop yields and global food security: will yield increase continue to feed the world?* Editorial: Australian Centre for International Agricultural Research, Canberra (ACIAR).

[24] FAO (Organización de las Naciones Unidas para la Agricultura y la Alimentación). *Buenas Prácticas Agrícolas para la Agricultura Familiar. Cadena de las Principales Hortalizas de Hojas en Argentina.* [Good Agricultural Practices for Family Farming. Chain of the Main Leaves Vegetables in Argentina.] Available online:

http://www.fcagr.unr.edu.ar/PHR/BPAparaHortalizasdeHoja202010.pdf. Accessed date: 12 March 2018.

[25] World Health Organization (WHO). Organización Panamericana de la Salud (OPS). (1993). División Salud y Ambiente. *Plaguicidas y Salud en las Américas*; OMS/OPS: Washington, DC, USA. [Pan American Health Organization (PAHO). (1993). Health and Environment Division. *Pesticides and Health in the Americas*]

[26] Bonansea, R. I., Filippi, I., Wunderlin, D. A., Marino, D. J. G. and Amé, M. V. (2018). The fate of glyphosate and AMPA in a freshwater endorheic basin: an ecotoxicological risk assessment. *Toxics*, 6 (3): 1-13.

[27] Cámara de Sanidad Agropecuaria y Fertilizantes (CASAFE), 2008. *Crece el volumen de uso de fitosanitarios, pero cae su toxicidad.* Available online: http://www.casafe.org/web_css/pop_up/comptox.htm. Accessed date: 12 March 2018. [*The volume of use of phytosanitary products grows, but its toxicity drops*.]

[28] Annett, R., Hamid, R. H. and Hontela, A. (2014). Impact of glyphosate and glyphosate-based herbicides on the freshwater environment. *Journal of Applied Toxicology*, 34: 458-479.

[29] Botta, F., Fauchon, N., Blanchoud, H., Chevreuil, M. and Guery, B. (2012). Phyt'Eaux Cités: application and validation of a programme to reduce surface water contamination with urban pesticides. *Chemosphere*, 86: 166-76.

[30] Coupe, R. H., Kalkhoff, S. J., Capel, P. D. and Gregoire, C. (2012). Fate and transport of glyphosate and aminomethylphosphonic acid in surface waters of agricultural basins. *Pest Management Science*, 68: 16-30.

[31] Grunewald, K., Wido Schmidt, C. U. and Gudrun, H. (2001). Behavior of glyphosate and aminomethylphosphonic acid (Ampa) in soils and water of reservoir Radeburg Ii Catchment (Saxony/Germany). *Journal of Plant Nutrition and Soil Science*, 164: 65-70.

[32] Mesnage, R. and Antoniou, M. N. (2018) Ignoring Adjuvant Toxicity Falsifies the Safety Profile of Commercial Pesticides. *Frontiers in Public Health*, 5: 361:1-8.

[33] Williams, G. M., Kroes, R. and Munro, I. C. (2000). Safety evaluation and risk assessment of the herbicide Roundup and its active ingredient, glyphosate, for humans. *Regulatory Toxicology and Pharmacology*, 31: 117–165.

[34] Solomon, K. R. and Thompson, D. G. (2003). Ecological risk assessment for aquatic organisms from over-water uses of glyphosate. *Journal of Toxicology and Environmental Health B. Critical Reviews*, 6: 289–324.

[35] Adam, A., Marzuki, A., Abdul, Rahman, H. and Abdul Aziz, M. (1997). The oral and intra-tracheal toxicities of ROUNDUP and its components to rats. *Veterinary and Human Toxicol*ogy, 39:147-151.

[36] Marc, J., Le Breton, M., Cormier, P., Morales, J., Belle, R. and Mulner-Lorillon, O. A. (2005). Glyphosate-based pesticide impinges on transcription. *Toxicology and Applied Pharmacology*, 203:1–8.

[37] Tsui, M. T. and Chu, L. M. (2003). Aquatic toxicity of glyphosate-based formulations: comparison between different organisms and the effects of environmental factors. *Chemosphere*, 52:1189–1197.

[38] Mottier, A., Kientz-Bouchart, V., Serpentini, A., Lebel, J. M., Jha, A. N. and Costil, K. (2013). Effects of glyphosate-based herbicides on embryo-larval development and metamorphosis in the Pacific oyster, *Crassostrea gigas*. *Aquatic Toxicology*, 128-129:67–78.

[39] Mottier, A., Pini, J. and Costil, K. (2014). Effects of a POEA surfactant system (Genamin T-200(R) on two life stages of the Pacific oyster, *Crassostrea gigas*. *The Journal of Toxicological Science*, 39: 211–215.

[40] Jayasumana, C., Fonseka, S., Fernando, A., Jayalath, K., Amarasinghe, M., Siribaddana, S., Gunatilake, S. and Paranagama, P. (2015). Phosphate fertilizer is a main source of arsenic in areas affected with chronic kidney disease of unknown etiology in Sri Lanka. *SpringerPlus* 4: 90.

[41] Benachour, N. and Seralini, G. E. (2009). Glyphosate formulations induce apoptosis and necrosis in human umbilical, embryonic, and placental cells. *Chemical Research in Toxicology*, 22: 97–105.

[42] Defarge, N., Takacs, E., Lozano, V. L., Mesnage, R., Spiroux de Vendomois, J., Seralini, G. E. and Szekacs, A. (2016). Co-formulants in glyphosate-based herbicides disrupt aromatase activity in human cells below toxic levels. *International Journal of Environmental Research* and *Public Health,* 13: 264.

[43] Nobels, I., Spanoghe, P., Haesaert, G., Robbens, J. and Blust, R. (2011) Toxicity ranking and toxic mode of action evaluation of commonly used agricultural adjuvants on the basis of bacterial gene expression profiles. *PLoS One*, 6 (11): 24139.

[44] Székács, I., Fejes, Á., Klátyik, S., Takács, E., Patkó, D., Pomóthy, J., Mörtl, M., Horváth, R., Madarász, E., Darvas, B. and Székács A. (2014). Environmental and toxicological impacts of glyphosate with its formulating adjuvant. *International Journal of Agricultural and Biological Engineering,* 8: 210–216.

[45] Zerin, T., Song, H., Gil, H. and Hong, S. (2017). Surfactant 4-nonylphenyl-polyethylene glycol stimulates reactive oxygen species generation and apoptosis in human neuroblastoma cells. *Journal of Environmental Sciences*, 53: 262–268.

[46] Gasnier, C., Laurant, C., Decroix-Laporte, C., Mesnage, R., Clair, E., Travert, C. and Seralini, G. (2011). Defined plant extracts can protect human cells against combined xenobiotic effects. *Journal of Occupational Medicine and Toxicology*, 6 (1). 3.

[47] Navarro, C. D. and Martinez, C. B. (2014). Effects of the surfactant polyoxyethylene amine (POEA) on genotoxic, biochemical and physiological parameters of the freshwater teleost *Prochilodus lineatus*. *Comparative Biochemistry and Physiology Part C*, 165: 83–90.

[48] Zhao, X.., Yang, G., Toyooka, T. and Ibuki, Y. (2015). New mechanism of gamma-H2AX generation: surfactant-induced actin disruption causes deoxyribonuclease I translocation to the nucleus and forms DNA double-strand breaks. *Mutation Research*, 794: 1-7.

[49] Mesnage, R., Defarge, N., Spiroux de Vendômois, J. and Séralini. G. E. (2015). Potential toxic effects of glyphosate and its commercial

formulations below regulatory limits. *Food and Chemical Toxicology,* 84:133-153.

[50] Mesnage, R., Renney, G., Séralini, G. E., Ward, M. and Antoniou, M. N. (2017). Multiomics reveal non-alcoholic fatty liver disease in rats following chronic exposure to an ultra-low dose of Roundup herbicide. *Scientific Report*, 7: 39328, 1-15.

[51] Niemeyer, J. C., de Santo, F. B., Guerra, N., Ricardo Filho, A. M. and Pech, T. M. (2018). Do recommended doses of glyphosate-based herbicides affect soil invertebrates? Field and laboratory screening tests to risk assessment. *Chemosphere*, 198: 154-160.

[52] European Chemicals Agency (ECHA). (2017). *Glyphosate not classified as a carcinogen by ECHA* (ECHA/PR/17/06, 2017). Available online: https://echa.europa.eu/glyphosate-not-classifiedas-a-carcinogen-by-echa. Accessed date: 26 March 2018.

[53] Newton, M., Howard, K. M., Kelpsas, B. R., Danhaus, R., Lottman, C. M. and Dubelman, S., (1984). Fate of glyphosate in an Oregon forest ecosystem. *Journal Agricultural and Food Chemistry*, 32: 1144-1151.

[54] Scribner, E. A., Battaglin, W. A., Gilliom, R. J. and Meyer, M. T. (2007). Concentrations of glyphosate, its degradation product, aminomethylphosphonic acid, and glufosinate in ground- and surface-water, rainfall, and soil samples collected in the United States, 2001-06. *U.S. Geological Survey Scientific Investigations Report 2007-5122*, pp. 1-111.

[55] Battaglin, W. A., Kolpin, D. W., Scribner, E. A., Kuivila, K. M. and Sandstrom, M. W. (2005). Glyphosate, other herbicides, and transformation products in Midwestern streams, 2002. *Journal of the American Water Resources Association*, 41: 323–332.

[56] Peruzzo, P. J., Porta, A. A., and Ronco, A. E. (2008). Levels of glyphosate in surface waters, sediments and soils associated with airect sowing soybean cultivation in north Pampasic Region of Argentina. *Environmental Pollution*, 156: 61–66.

[57] Aparicio, V. C., De Gerónimo, E., Marino, D., Primost, J., Carriquiriborde, P. and Costa, J. L. (2013). Environmental fate of

glyphosate and aminomethylphosphonic acid in surface waters and soil of agricultural basins. *Chemosphere*, 93: 1866–1873.

[58] Ronco, A. E., Marino, D., Abelando, M., Almada, P. and Apartin, C.D. (2016). Water quality of the main tributaries of the Parana basin: glyphosate and ampa in surface water and bottom sediments. *Environmental Monitoring and Assessment,* 188 (8): 458.

[59] Primost, J. E., Marino, D., Aparicio, V. C., Costa, J. L. and Carriquiriborde, P. (2017). Glyphosate and ampa, "pseudo-persistent" pollutants under real-world agricultural management practices in the mesopotamic pampas agroecosystem, Argentina. *Environmental Pollution*, 229: 771-779.

[60] Maqueda, C., Undabeytia, T., Villaverde, J. and Morillo, E. (2017). Behaviour of glyphosate in a reservoir and the surrounding agricultural soils. *Science of the Total Environment*, 1 (593-594):787-795.

[61] Grandcoin, A., Piel, S. and Baurès, E. (2017). Amino-methylphosphonic acid (AMPA) in natural waters: Its sources, behavior and environmental fate. *Water Research* 15: 117:187-197.

[62] Poiger, T., Buerge, I. J., Bächli, A., Müller, M. D. and Balmer, M. E. (2017). Occurrence of the herbicide glyphosate and its metabolite AMPA in surface waters in Switzerland determined with on-line solid phase extraction LC-MS/MS. *Environmental Science and Pollution Research International*, 24: 1588-1596.

[63] Lupi, L., Miglioranza, K. S. B., Aparicio, V. C., Marino, D., Bedmar, F. and Wunderlin, D. A. (2015). Occurrence of glyphosate and Ampa in an agricultural watershed from the southeastern region of Argentina. *Environmental Science and Technology*, 536: 687-694.

[64] Okada, E., Costa, J. L. and Bedmar, F. (2016). Adsorption and mobility of glyphosate in different soils under no-till and conventional tillage. *Geoderma*, 263: 78-85.

[65] Vereecken, H. (2005). Mobility and leaching of glyphosate: a review. *Pest Management Science*, 61: 1139 1151.

[66] Kjær, J., Ernsten, V., Jacobsen, O. H., Hansen, N., de Jonge, L. W. and Olsen, P. (2011). Transport modes and pathways of the strongly

sorbing pesticides glyphosate and pendimethalin through structured drained soils. *Chemosphere*, 84: 471-479.

[67] Skark, C., Zullei-Seibert, N., Schottler, U. and Schlett, C. (1998). The occurrence of glyphosate in surface water. *International Journal of Environmental Analytical Chemistry*, 70: 93–104.

[68] Mörtl, M., Németh, G., Juracsek, J., Darvas, B., Kamp, L., Rubio, F. and Székács, A. (2013). Determination of glyphosate residues in Hungarian water samples by immunoassay. *Microchemical Journal*, 107: 143-151.

[69] Sanchís, J., Kantiani, L., Llorca, M., Rubio, F., Ginebreda, A., Fraile, J., Garrido, T., and Farré, M. (2012). Determination of glyphosate in groundwater samples using an ultrasensitive immunoassay and confirmation by on-line solid-phase extraction followed by liquid chromatography coupled to tandem mass spectrometry. *Analytical and Bioanalytical Chemistry*, 402: 2335-2345.

[70] Villeneuve, A., Larroudé, S. and Humbert, J. F. (2011). Herbicide contamination of freshwater ecosystems: impact on microbial communities. In: Stoytcheva, M. (Ed.), *Pesticides formulations, effects, fate.* pp. 285-312. Available online: https://www.intechopen.com/books/pesticides-formulations-effects-fate/herbicide-contamination-of-freshwater-ecosystemsimpact-on-microbial-communities. Accessed date: 4 March 2018.

[71] Rosenbom, A. E., Brüsch, W., Juhler, R. K., Ernstsen, V., Gudmundsson, L., Kjær, J., Plauborg, F., Grant, R., Nyegaard, P. and Olsen, P. (2010). *The Danish pesticide leaching assessment programme monitoring results May 1999–June 2009*. Geological Survey of Denmark and Greenland Ministry of Climate and Energy and Faculty of Agricultural Sciences. Available online: http://pesticidvarsling.dk/xpdf/vap-results-99-09.pdf. Accessed date: 4 March 2018.

[72] Zhang, C., Feng, L., He, T. T., Yang, C. H., Chen, G. Q. and Tian, X. S. (2015). Investigating the mechanisms of glyphosate resistance

in goosegrass (*Eleusine indica*) population from South China. *Journal of Integrative Agriculture*, 14: 909-918.

[73] CODEX Alimentarius. (2013). *Pesticide residues in food and feed: 158 Glyphosate.* Food and Agriculture Organization and World Health Organization. Available online: http://ww.codexalimentarius.net/pestres/data/pesticides/details.html?id=158, Accessed date: 9 March 2018.

[74] EPA (Environmental Protection Agency). (2013). Electronic code of federal regulations, title 40: protection of environment. PART 180-*Tolerances and Exemptions for Pesticide Chemical Residues in Food. Subpart C-Specific Tolerances.* Environmental Protection Agency, Washington, DC. Available online: http://www.ecfr.gov/cgi-bin/text-idx?tpl=/ecfrbrowse/Title40/40cfr180_main_02.tpl, Accessed date: 9 March 2018.

[75] Arregui, M. C., Lenardón, A., Sanchez, D., Maitre, M. I. and Enrique, S. (2004). Monitoring glyphosate residues in transgenic glyphosate-resistant soybean. *Pest Management Science*, 60: 163-166.

[76] Bøhn, T., Cuhra, M., Travail, T., Sanden, M., Fagan, J. and Primicerio, R. (2014). Compositional differences in soybeans on the market: glyphosate accumulates in Roundup Ready GM soybeans. *Food Chemistry*, 153: 207-215.

[77] Cuhra, M. (2015). *Review of GMO safety assessment studies: glyphosate residues in Roundup Ready crops is an ignored issue.* Environmental Science Europe 27 (20). Available online: https://doi.org/10.1186/s12302-015-0052-7. Accessed date: 9 March 2018.

[78] Çetin, E., Şahan, S., Ülgen, A. and Şahin, U. (2017). DLLME-spectrophotometric determination of glyphosate residue in legumes. *Food Chemistry,* 230: 567-571.

[79] Niemann, L., Sieke, C., Pfeil, R. and Solecki, R. (2015). A critical review of glyphosate findings in human urine samples and comparison with the exposure of operators and consumers. *Journal for Consumer Protection and Food Safety*, 10: 3–12.

[80] von Soosten, D., Meyer, U., Hüther, L., Dänicke, S., Lahrssen-Wiederholt, M., Schafft, H., Spolders, M. and Breves, G. (2016). Excretion pathways and ruminal disappearance of glyphosate and its degradation product aminomethylphosphonic acid in dairy cows. *Journal of Dairy Science*, 99: 5318-5324.

[81] Krüger, M., Schledorn, P., Schrödl, W., Hoppe, H. W., Lutz, W. and Shehata, A. A. (2014a). Detection of glyphosate residues in animals and humans. *Journal of Environmental and Analytical Toxicology*, 4: 2. Available online: https://doi.org/10.4172/2161-0525.1000210. Accessed date: 9 March 2018.

[82] Krüger, M., Schrödl, W., Pedersen, I., Schledorn, P. and Shehata, A. A. (2014b). Detection of glyphosate in malformed piglets. *Journal of Environmental and Analytical Toxicology,* 4: 5. Available online: https://doi.org/10.4172/2161-0525.1000230. Accessed date: 9 March 2018.

[83] WHO (World Health Organization), 2005. *Glyphosate and AMPA in Drinking-Water.* Available online: http://www.who.int/water_sanitation_health/dwq/chemicals/glyphosateampa290605.pdf, Accessed date: 9 March 2018.

[84] Vandenberg, L. N., Blumberg, B., Antoniou, M. N., Benbrook, C. M., Carroll, N., Colborn, T., Everett, L. G., Hansen, M., Landrigan, P. J., Lanphear, B. P., Mesnage, R., vom Saal, F. S., Welshons, W. V. and Myers, J. P. (2017). Is it time to reassess current safety standards for glyphosate-based herbicides? *Journal of Epidemiology and Community Health,* 71: 613–618.

[85] Mamy, L., Barriuso, E. and Benoît, G. (2016). Glyphosate fate in soils when arriving in plant residues. *Chemosphere*, 154: 425-433.

[86] Imfeld, G. and Vuilleumier, S. (2012). Measuring the effects of pesticides on bacterial communities in soil: A critical review. *European Journal of Soil Biology*, 49: 22–30.

[87] Jacobsen, C. S. and Hjelmsø, M. H. (2014). Agricultural soils, pesticides and microbial diversity. *Current Opinion in Biotechnology,* 27: 15-20.

[88] Reddy K., Abbas, H., Zablotowicz, R., Abel, C. and Koger C. (2007). Mycotoxin occurrence and *Aspergillus flavus* soil propagules in a corn and cotton glyphosate-resistant cropping systems. *Food Additives and Contaminants*, 24: 1367-1373.

[89] Krzyśko-Łupicka, T. and Sudol, T. (2008). Interactions between glyphosate and autochthonous soil fungi surviving in aqueous solution of glyphosate. *Chemosphere*, 71: 1386-1391.

[90] Ratcliff, A., Matt, D., Busse, M. and Shestak, C. (2006). Changes in microbial community structure following herbicide (glyphosate) additions to forest soils. *Applied Soil Ecology*, 34: 114-124.

[91] Zabaloy, M., Garland, J. and Gómez, M. (2008). An integrated approach to evaluate the impacts of the herbicides glyphosate, 2,4-D and metsulfuron-methyl on soil microbial communities in the Pampas region, Argentina. *Applied Soil Ecology*, 40: 1‑12.

[92] Mijangos, I., Becerril, J., Albizu, I., Epelde, L. and Garbisu, C. (2009). Effects of glyphosate on rhizosphere soil microbial communities under two different plant compositions by cultivation-dependent and -independent methodologies. *Soil Biology and Biochemistry*, 41: 505–513.

[93] Carlisle, S. M. and Trevors, J. T. (1988). Glyphosate in the environment. *Water, Air and Soil Pollution*, 39: 409-420.

[94] Dick, R. E. and Quinn, J. P. (1995). Glyphosate-degrading isolates from environmental samples: occurrence and pathways of degradation. *Applied Microbiology and Biotechnology*, 43: 545–550.

[95] Kremer, R. J., Means, N. E. and Kim, S. (2005). Glyphosate affects soybean root exudation and rhizosphere micro-organisms. *International Journal of Environmental Analytical Chemistry*, 85: 1165-1174.

[96] Liu, L., Punja, Z. K. and Rahe, J. E. (1997). Altered root exudation and suppression of induced lignification as mechanisms of predisposition by glyphosate of bean roots (*Phaseolus vulgaris L.*) to colonization by *Pythium* spp. *Physiological and Molecular Plant Pathology*, 51: 111-127.

[97] Kuklinsky-Sobral, J., Araujo, W. L., Mendes, R., Pizzirani-Kleiner, A. A. and Azevedo, J. L. (2005). Isolation and characterization of endophytic bacteria from soybean (*Glycine max*) grown in soil treated with glyphosate herbicide. *Plant and Soil*, 273: 91–99.

[98] Busse, M. D., Ratcliff, A. W., Shestak, C. J. and Powers, R. F. (2001). Glyphosate toxicity and the effects of long-term vegetation control on soil microbial communities. *Soil Biology and Biochemistry*, 33: 1777-1789.

[99] Motavalli, P. P., Kremer, R. J., Fang, M. and Means, N. E. (2004). Impact of genetically modified crops and their management on soil microbially mediated plant nutrient transformations. *Journal of Environmental Quality*, 33: 816–824.

[100] Imparato, V., Santos, S. S., Johansen, A., Geisen, S. and Winding, A. (2016). Stimulation of bacteria and protists in rhizosphere of glyphosate-treated barley. *Applied Soil Ecology*, 98: 47–55.

[101] Druille, M., García-Parisi, P. A., Golluscio, R. A., Cavagnaro, F. P. and Omacini, M. (2016). Repeated annual glyphosate applications may impair beneficial soil microorganisms in temperate grassland. *Agriculture, Ecosystems and Environment,* 230: 184-190.

[102] Newman, M., Hoilett, N., Lorenz, N., Dick, R., Liles, M., Ramsier, C. and Kloepper, J. (2016). Glyphosate effects on soil rhizosphere-associated bacterial communities. *Science of the Total Environment*, 543: 155-160.

[103] Dennis, P., Kukulies, T., Forstner, C., Orton, T. and Pattison, A. (2018). The efects of glyphosate, glufosinate, paraquat and paraquat-diquat on soil microbial activity and bacterial, archaeal and nematode diversity. *Scientific Report*, 8: 2119.

[104] Muturi, E., Donthu, R., Fields, C., Moise, I. and Kim, C. H. (2017). Effect of pesticides on microbial communities in container aquatic habitats. *Scientific Reports*, 7: 44565.

[105] Al-Rajab, A. J. and Schiavon, M. (2010). Degradation of 14C-glyphosate and aminomethylphosphonic acid (AMPA) in three agricultural soils. *Journal of Environmental Sciences*, 22: 1374–1380.

[106] Sposito, G. (2008). *The chemistry of soils*. Oxford University Press, New York, Chapter 1: 3-25.

[107] Cheng, H. Y. (1990). Pesticides in the Soil Environment-An Overview. Chapter 1. En Pesticides in the Soil Environment: Processes, Impacts and Modeling, SSSA Book Series 2. Ed. H. H. Cheng. *Soil Science Society of America*.

[108] Rueppel, M. L., Brightwell, B. B., Schaefer, J. and Marvel, J. (1977). Metabolism and degradation of glyphosate in soil and water. *Journal of Agricultural and Food Chemistry,* 25:517–528.

[109] Landry, D., Dousset, S., Fournier, J. C. and Andreux, F. (2005). Leaching of glyphosate and AMPA under two soil management practices in Burgundy vineyards (Vosne-Roman'ee, 21- France). *Environmental Pollution,* 138:191–200.

[110] Degenhardt, D., Humphries, D., Cessna, A. J., Messing, P., Badiou, P. H., Raina R., Farenhorst, A. and Pennock, D. J. (2012). Dissipation of glyphosate and aminomethylphosphonic acid in water and sediment of two Canadian prairie wetlands. *Journal of Environmental Science, Pesticides, Food Contaminants, and Agricultural Wastes,* 47: 631-639.

[111] Comerford, N. (2003). *Soil: Largest reactor on the planet?* Available on: https://soilsmatter.wordpress.com/2013/12/02/soil-largest-reactor on-the planet. Accessed date. 9 March 2018.

[112] Schwarzenbach, R. P., Gschwend, P. M. and Imboden, D. M. (1993). *Environmental Organic Chemistry*. John Wiley & Sons, Inc., New York, NY, 681 pp.

[113] Borggaard, O. K. and Elberling, B. (2004). *Pedological Biogeochemistry*. Paritas, Brøndby, Denmark, 483 pp.

[114] Sheals, J., Sjoberg, S. and Persson, P. (2002). Adsorption of glyphosate on goethite: molecular characterization of surface complexes. *Environmental Science and Technology,* 36:3090–3095.

[115] Barja, B. C. and dos Santos Afonso, M. (2005). Aminomethyl-phosphonic acid and glyphosate adsorption onto goethite: a comparative study. *Environmental Science Technology,* 39:585-592.

[116] Jlassi, K., Chehimi, M. M. and Thomas, S. (2017). *Clay-Polymer Nanocomposites.* Elsevier Inc, Chapter 1: 1-25. ISBN: 978-0-323-46153-5.

[117] Abend, S. and Lagaly, G. (2000). Sol–gel transitions of sodium montmorillonite dispersions. *Applied Clay Science,* 16: 201–227.

[118] Mahkam, M., Davatgar, M., Rezvani, Z. and Nejati, K. (2013) Preparation and pH-sensitive polymers/layered double hydroxide hybrid beads for controlled release of insulin. *International Journal of Polymeric Materials and Polymeric Biomaterials,* 62: 57-60.

[119] Kakati, K., Pugazhenthi, G. and Iyer, P. K. (2011). Effect of Organomodified Ni-Al Layered Double Hydroxide (OLDH) on the Properties of Polypropylene (PP)/LDH Nanocomposites. *International Journal of Polymeric Materials and Polymeric Biomaterials,* 61: 931–948.

[120] Shoval, S. and Yariv, S. (1979). The interactions between Roundup (Glyphosate) and montmorillonite. Part I Infrared study of the sorption of glyphosate by montmorillonite. *Clay and Clay Minerals*, 27: 19-28.

[121] de Jonge, H., de Jonge, L. W., Jacobsen O. H., Yamaguchi, T. and Moldrup, P. (2001). Glyphosate sorption in soils of different pH and phosphorus content. *Soil Science,* 166: 230-238.

[122] Damonte, M., Torres Sánchez, R. M. and dos Santos Afonso, M. (2007). Some aspects of the glyphosate adsorption on montmorillonite and its calcined form. *Applied Clay Science,* 36:86–94.

[123] Glass, R. L. (1987). Adsorption of glyphosate by soils and clay minerals. *Journal Agricultural Food Chemistry,* 35:497–500.

[124] McConnell, J. S. and Hossner, L. R. (1985). pH Dependent adsorption isotherms of Glyphosate. *Journal Agricultural Food Chemistry,* 33:1075-1078.

[125] Gimsing, A. L. and dos Santos Afonso, M. (2005). Glyphosate. Biogeochemistry of Chelating Agents. *American Chemical Society*, Chapter 16: 263-277.

[126] Gimsing, A. L. and Borggaard, O. K. (2002). Competitive adsorption and desorption of glyphosate and phosphate on clay silicates and oxides. *Clay Minerals,* 37:509–515.

[127] Morillo, E., Undabeytia, T. and Maquada, C. (1997). Adsorption of glyphosate on the clay mineral montmorillonite: effect of Cu(II) in solution and adsorbed on the mineral. *Environmental Science Technology,* 31:3588–3592.

[128] Borggaard, O. K. and Gimsing, A. L. (2008). Fate of glyphosate in soil and the possibility of leaching to ground and surface waters: a review. *Pest Management Science,* 64:441–456.

[129] Gimsing, A. L., Borggaard, O. K. and Bang, M. (2004). Influence of soil composition on adsorption of glyphosate and phosphate by contrasting Danish surface soils. *European Journal Soil Science*, 55: 183–191.

[130] Vereecken, H. (2005). Mobility and leaching of glyphosate: a review. *Pest Management Science,* 61: 1139–1151.

[131] Piccolo, A. G. and Celano and Conte, P. (1996). Adsorption of glyphosate by humic substances. *Journal of Agricultural and Food Chemistry,* 44: 2442-2446.

[132] Dion, H. M., Harsh, J. B. and Hill, H. H. (2001). Competitive sorption between glyphosate and inorganic phosphate on clay minerals and low organic matter soils. *Journal Radioanalytical Nuclear Chemistry,* 249: 385–390.

[133] Wang, Y. J., Zhou, D. M. and Sun, R. J. (2005). Effects of phosphate on the adsorption of glyphosate on three different types of Chinese soils. *Journal Environmental Science,* 17: 711–715.

In: Glyphosate
Editor: Ron Quinn

ISBN: 978-1-53614-002-6

Chapter 2

SLOW RELEASE OF GLYPHOSATE FROM LDH-GLYPHOSATE HYBRID COMPOUNDS IN AQUEOUS SOLUTIONS OF DIFFERENT ANIONS

Emanoel Hottes, André Marques dos Santos, Cristiane Martins Cardoso, Rosane Nora Castro and Marcelo Hawrylak Herbst**

Departamento de Química,
Universidade Federal Rural do Rio de Janeiro,
Seropédica, RJ, Brazil

ABSTRACT

The objective of this work was to investigate the best way to achieve slow release of glyphosate using layered double hydroxides (LDHs), to minimize the environmental impacts of this herbicide. Understanding the release dynamics of glyphosate in homogeneous solutions of anions that

---

* Corresponding Author Email: herbst_marcelo@yahoo.com.br.

are commonly found in the soil is important, since many crops require prior soil correction, altering the pH and also the availability of anions. LDHs are best stabilized by anions with high charge density, such as carbonate. In turn, glyphosate is a potential intercalant due to its charge density. At pH 9, glyphosate is deprotonated both in its carboxylate and phosphonate portions, so it can be intercalated to form a hybrid material with unique properties. The LDH used in this study was prepared by a coprecipitation process with variable pH, and was characterized by infrared spectroscopy, X-ray powder diffraction and ^{27}Al and ^{31}P solid-state nuclear magnetic resonance. The release kinetics of glyphosate was analyzed in solutions containing carbonate, nitrate and bicarbonate anions, either by the colorimetric method of Ruhemann and by HPLC. Both detection methods indicated that the carbonate anions promoted greater release of glyphosate.

Keywords: glyphosate, layered double hydroxides, slow-release matrix

INTRODUCTION

Glyphosate, or N-(phosphonomethyl)glycine, mainly sold under the trade name Roundup®, is a herbicide belonging to the class of phosphorylated amino acids (Acquavella et al., 2004). The most common commercial formula of glyphosate, produced by Monsanto, is composed of 359 $g.L^{-1}$ (acid equivalent) of isopropylamine salt plus a surfactant blend. It is important to note that the word glyphosate is used to refer both to the acid and its salts, since they are biologically equivalent (Franz, 1985; Quinn, 1993).

The use of glyphosate is indicated to control annual or perennial weeds, both monocots and dicots, in crops such as irrigated rice, citrus, apple, coffee, sugarcane, soybean and tobacco, as well as forage grasses, among others. Glyphosate is a highly potent herbicide due to its excellent ability to translocate in plants. When it is applied, part is directly absorbed by the plants while another portion goes directly to the soil, where it is biodegraded by microorganisms (Giesy et al., 2000).

With respect to the environmental impacts of agrochemicals, ideally they should remain active for a determined period and then disappear by degradation. However, this pattern does not occur with glyphosate. A portion does not reach the target weed, and is therefore carried to other sites where it tends to be deposited, therefore being susceptible to various chemical and physical processes, such as complexation with cations present in the soil, biodegradation, leaching and percolation (Rueppel et al., 1977).

Among such processes, those of greater concern regarding glyphosate are complexation with heavy metals present in the soil and water, such as Ca^{2+}, Mg^{2+}, Ni^{2+}, Fe^{+2}, Fe^{+3} and Al^{+3}; sorption with sediments and particles suspended in soil and water; absorption and metabolism by plants; and biodegradation by soil microorganisms (Giesy et al., 2000; Hensley, 1978; Glass, 1987; Hoagland, 1979).

A key aspect that influences the mobility of glyphosate in soil and water is the pH of the medium, since the compound has distinct forms that accompany variations of this physico-chemical property. At pH values lower than 0.8, practically all the glyphosate is found in the form of completely protonated molecules. In solutions with pH values in the range of 7 to 8, the predominant form of glyphosate is gly^{2-}, with only the phosphonate group protonated. It was initially believed that at that pH value, the amine portion of the molecule would hold the proton, but theoretical studies carried out by Peixoto and colleagues (2015) and confirmed experimentally by Liu and colleagues (2016) showed that the protonation occurs in the phosphonate group portion. Finally, at pH values higher than 9, glyphosate is found in its gly^{3-} anion form.

Action of Glyphosate

When penetrating plants, glyphosate is carried by the phloem and is rapidly translocated to all parts, although tending to accumulate mainly in the meristematic regions (Giesy et al., 2000; Haderlie et al., 1978). Sometime after application of glyphosate, leaf chlorosis occurs, followed by necrosis, as well as other symptoms, like wrinkling or malformation (mainly in sprouting regions), meristematic necrosis, also affecting rhizomes and runners of perennial plants (Campbell et al., 1976). The action mechanism of glyphosate in plants is very particular, because it is the only herbicide able to inhibit the enzyme 5-enolpyruvylshikimate-3-phosphate-synthase (EPSPs), which is directly involved in synthesizing shikimic acid and pyruvate phosphate synthase, which in turn are associated with the production of three amino acids that are essential to plants: tryptophan, phenylalanine and tyrosine (Ashton, 1981; Jaworski, 1972). Interrupting the production of these aromatic amino acids by plants is a good strategy for herbicidal action, since the shikimic pathway is only present in plants, i.e., it is not found in animals (Duke et al., 2008).

Action of Glyphosate in Soil and Water and Its Transformation Processes

Typically, the lower the solubility of molecules is in water, the greater will be their capacity for sorption in the soil. However, glyphosate is an exception to this rule, because according to the pH of the medium, the molecule has very high ability to dissolve in water and also high potential for sorption in soil (Miles & Moye, 1998; McConnell & Hossner, 1985). The high potential for sorption in soil attributed to glyphosate can be associated with the interaction of van de Waals forces, the formation of hydrogen bonds with the humic compounds present in soil, and also the formation of complex structures by establishment of covalent bonds with the metal oxide ions present in soil. The last of these mechanisms is the most relevant for oxidic soils, so that glyphosate tends to remain in the medium as a bound residue, a factor that can influence the bio-transformation process by hindering the availability of molecules to microorganisms present in the soil (Giesy et al., 2000; Prata et al., 2005; Piccolo et al., 1996).

A large range of active herbicidal molecules can undergo chemical transformation by the action of soil microorganisms, as already mentioned, as is the case of glyphosate. Compared to its metabolites, glyphosate is not highly toxic, since its half-life is relatively short, from a few days to a few months depending on the levels of clay and organic materials present in the soil, as well as the metabolic rate of the microorganisms. This rate, in turn, varies from one soil type to another (Rueppel et al., 1977; Araújo et al., 2003; Glass, 1987; Hensley, 1978), but it is known that 50% of the molecules can be metabolized in an average period of 28 days, reaching 90% in approximately 90 days. For that reason, metabolites or degradation products can be easily identified, as is the case of AMPA. This compound is a degradation product of glyphosate via the action of the soil microbiota (Giesy et al., 2000; Araújo et al., 2003).

The degradation process of glyphosate in itself is relatively fast, because it is performed by a large number of microorganisms that use that compound as a source of energy, phosphorus, nitrogen and carbon through metabolic pathways, one of which results in AMPA and the other sarcosine. Unlike glyphosate, AMPA has a very long half-life, which can surpass 240 days, with degradation by carbon dioxide and ammonia (Battaglin et al., 2005; Araújo et al., 2003).

In water, glyphosate mainly disperses by microbiological degradation and also by union with sediments, although this second mechanism does not occur quickly like in the soil. However, the formation of AMPA also occurs due to the action of the aquatic microflora (Dick & Quinn, 1995; Gomez & Sagardoy, 1985).

Slow Release of Agrochemicals

Interest has been growing in recent years to ascertain the effects of agrochemicals on the environment. Agrochemicals are often applied in doses that exceed the level necessary to achieve optimal control of a determined target weed, disease or pest. A process that helps reduce the loss of these chemicals is related to the physical event of adsorption by the soil matrix, as mentioned before. That process is initially superficial, but is followed by a period of slow diffusion of the herbicide in the soil aggregates. However, if rain occurs relatively soon after application of herbicides in the soil, they tend to move freely by leaching, enabling them to accumulate in groundwater and also to be carried by runoff to other areas, with little or no contact with the adsorbent sites in the soil (Barja & Afonso, 2005).

Knowledge of the adsorption processes of the various matrixes in the soil is the base to understand and design controlled release vehicles. These have attracted strong interest by herbicide makers, since these formulations tend to mitigate the deleterious effects of these compounds on the environment and also the farmworkers involved in handling and applying them. In the formulations of substances adsorbed in matrixes, only a small part of the active ingredient will be available immediately in the medium, with the remainder being encapsulated in the inert matrix. This portion will be released gradually, at a rate depending on the specific controlled release mechanism. However, the release process is also closely associated with the nature of the adsorbed molecule used for the controlled release as well as by the possible existence of exchange agents in the medium.

Layered Double Hydroxides

Minerals belonging to the anionic clay family have been described in the literature by mineralogists for many decades. These materials are relatively rare in nature, but can easily be synthesized in the laboratory (Miyata, 1983). The anionic clays are mainly associated with metamorphic formations, and can also occur in saline deposits. This indicates their formation does not depend on extreme physical factors like high temperature and pressure (Yamamoto et al., 1995).

In general, the term anionic clay is used to designate natural or synthetic layered double hydroxides, which have negatively charged species in their interlayer spaces, as presented in Figure 1. The normally used expression “layered double hydroxide”, or “LDH”, refers to the presence of two metal cations in the lamellar layer of these compounds. One of the best known LDHs is hydrotalcite, a natural clay that contains carbonate anions in the interlamellar region, which in turn is composed of magnesium and aluminum, with a basically two-dimensional structure (Constantino & Pinnavaia, 1994).

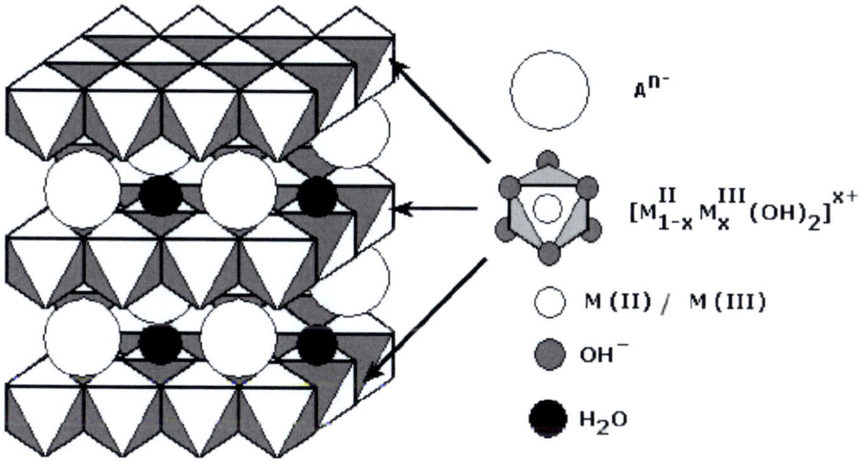

Figure 1. Scheme of the structural organization of LDH (Adapted from Constantino & Pinnavaia, 1995).

Structure of LDHs

Layered double hydroxides have a basic structure like that of brucite, a mineral composed of magnesium hydroxide where the Mg^{2+} cation is octahedrally coordinated by six hydroxyl anions. These octahedrons share edges, forming lamellae or layers (Scarpellini et al., 2015). These layers tend to be stacked on each other, stabilized and joined by interactions with agents present in their interlamellar domain (Crepaldi et al., 2000). What occurs in the structure of LDHs is that part of the Mg^{2+} ions are isomorphically substituted by trivalent cations such as Al^{3+}, Fe^{3+}, Ga^{3+}, thus giving the layer a positive charge (Crepaldi & Valim, 1998). Therefore, to maintain the electroneutrality of the system as a whole, anionic molecules together with water molecules must be present inside, resulting in a hydrotalcite structure (Kostadinova et al., 2016).

The general formula of LDHs is as follows:

$[M_{1-x}(II)M_x(III)(OH_2)]^{x+}A^{m-}{}_{x/m} \cdot nH_2O$

where:

- M(II) represents a divalent cation, which can be: Mg^{2+}, Zn^{2+}, Ca^{2+}, Mn^{2+}, Ni^{2+};
- M(III) corresponds to a trivalent cation, which can be: Al^{3+}, Co^{3+}, Fe^{3+}, Mn^{3+}, Ga^{3+};
- A^{m-} represents an anion with m^- charge, such as: CO_3^{2-}, Cl^-, NO_3^- and various other anionic organic compounds (Crepaldi et al., 2002).

The metal cations used to synthesize LDHs should have in common octahedral coordination and ionic radius between 0.5 and 0.74 Å for the structure to be formed. However, according to Theiss et al., (2016), cations like Ca^{2+} that have an ionic radius greater than that range can also be structural constituents of LDHs, as is the case of hydrocalumite.

The ratio between the divalent and trivalent cations can vary from 1 to 6, which corresponds to a range between 0.14 and 0.5 in the parameter x of the general formula. That ratio determines the charge density of the compound's layer, also influencing the crystallinity and ion exchange capacity of the material (Crepaldi et al., 2002). Ratios outside this range tend to produce individual metal hydroxides. The main result of a reduction of the M^{2+}/M^{3+} ratio is an increase in the material's charge density, which in turn alters the capacity for exchange and intercalation of anions (Yang et al., 2002).

A very important factor in preparing LDHs is the capacity of the intercalated anions to stabilize the lamellar structure. The greater this stabilization capacity is, the larger will be the quantity of material formed (Crepaldi et al., 2002). A wide range of anions, both organic and inorganic, can be intercalated in the interlamellar regions, with varied sizes. There is no restriction on the nature of the anions that will offset the positive charge of the lamellae. The basal spacing of the material will be determined by the number, size, orientation and strength of the link between the anions and hydroxyl groups of the layers (Prinetto et al., 2000).

Because of the stabilization ability of the anions, the greater their negative charge density is, the greater will be their electrostatic interaction with the lamellar layer. However, when organic anions are present, the interactions of the carbon chains will have an influence, so analyses considering only the concepts of charge cannot be applied alone (Crepaldi et al., 2002).

The inorganic anions follow the order indicated below regarding lamellar stabilization capacity:

$$CO_3^{2-} > OH^- > F^- > Cl^- > Br^- > NO_3^- > I^-$$

Analysis of this order reveals that LDHs that contain carbonate as the anions will provide the greatest stability, with these being relatively inefficient in the ion exchange process (Crepaldi & Valim, 1998). On the other hand, LDHs that contain anions at the opposite extreme of the series will be the most unstable and have the greatest tendency to exchange their anions. An explanation is that divalent anions have greater affinity for the lamellae due to the electrostatic interactions, compared to monovalent ones (Reichle, 1986).

Many methods are described in the literature to synthesize LDH, both direct and indirect. Studies have proved that properties like crystallinity of the material are directly influenced by the method employed in their synthesis. However, the most practical and efficient method that is widely used is coprecipitation at variable pH (Carrado et al., 1988). That method consists of adding a solution containing the salts of the divalent and trivalent cations to a hydroxide solution containing the anions targeted for intercalation. Some precautions should be taken in the synthesis process with variable pH regarding the proportion of reagents, reaction volume and temperature, agitation as well as the pH, to obtain a final product with good properties and purity (Pushparaj et al., 2015)

EXPERIMENTAL PROCEDURE

Materials

All the reagents used in the experiments described herein, except glyphosate, which was previously treated, had analytical purity degree and were used without previous treatment. Deionized water was used in the release assays. The glyphosate utilized was purified by treating an aqueous solution of Roundup WG (Monsanto, in the form of ammonium salt), purchased commercially, with excess acid and subsequently recrystallized in deionized water. This reaction produced an average yield of 70% of the initial weight of the Roundup used.

Synthesis of the LDH in the Mg-Al-Glyphosate System

The double hydroxide in the Mg-Al-glyphosate system was synthesized by the direct coprecipitation method with variable pH by slow addition of a solution containing 0.224 mol of $Mg(NO_3)_2.6H_2O$ and 0.112 mol of $Al(NO_3)_3.9H_2O$ into a solution containing 0.224 mol of glyphosate and 1.12 mol of NaOH, followed by adjustment of solution's pH to 9 in the case of the Mg-Al-Gly material, with this solution being submitted to thermal treatment via reflux for 24 hours at approximately 80 °C with vigorous agitation. After the thermal treatment, the suspension was centrifuged and the solid was then washed with deionized water and dried in an oven at 80°C for 8 hours (Meng et al., 2005).

Release Tests

The release kinetics tests were conducted by adapting methods described in the literature. Initially 25 mL samples were prepared of solutions containing 0.1, 0.4, 0.8, 1.5 or 2.5 mmol of the following sodium salts: nitrate, bicarbonate and carbonate. Then 100 mg of LDH prepared as described in the previous item was added separately to these solutions and left in contact for 2 hours, under constant agitation on an orbital table shaker at room temperature. Next, the suspensions formed were centrifuged and the supernatant was set aside for subsequent quantification of glyphosate (Meng et al., 2005).

Characterization of the Compounds and Release Testing

To study the structural properties of the LDHs synthesized in the Mg-Al-glyphosate system, conventional techniques were employed, such as X-ray powder diffraction, infrared spectroscopy and solid-state nuclear magnetic resonance spectrometry.

The XRD analyses of the compounds were carried out with a Rigaku Ultima IV diffractometer, using a graphite monochromator, nickel filter and CuKα radiation (1.54 angstroms), employing a scanning range (2θ) from 5° to 80° and step of 0.02°/2θ per minute. The objective of the XRD analyses was to identify and measure the possible phases or crystalline phase and the structural properties of the lamellar materials. By examining the diffractograms it was possible to characterize the compounds as being lamellar and determine the basal spacing and crystallinity degree.

The measurements in the infrared region were obtained using attenuated reflectance in an air atmosphere, to analyze the structural aspects of the synthesized compounds. The spectroscopic measures in the mid-infrared region (400 – 4000 cm^{-1}) were conducted with a Bruker Vertex 70 FTIR-ATR spectrometer with spectral resolution of 4 cm^{-1} and 64 scans. Spectroscopy in the mid-infrared region was used to identify the functional groups and anions intercalated in the LDHs. This method involves analyzing the vibrations of the symmetric octahedral unit of the LDHs, besides those of the hydroxyl groups and intercalated anions. The bands obtained were compared with data from the literature by analyzing the position of the bands, allowing identification of the functional groups as well as the transformations in them by absorption of the anions of interest.

The hybrid materials obtained were analyzed by solid-state nuclear magnetic resonance (NMR) spectroscopy with a Bruker DPX 400 spectrometer (400 MHz for ^{1}H) with a 4 mm HRMAS (High Resolution Magic Angle Spinning) probe.

The solutions obtained in the release test were analyzed by two methods: the colorimetric method of Ruhemann and high-performance liquid chromatography (HPLC).

The colorimetric method with Ruhemann's purple is widely used to quantify amino acids. Since glyphosate has the typical structure and behavior of an amino acid, this method is also used for its quantification, with some necessary adaptations as described by Bhaskara and Nagaraja (2006). The purple complex of Ruhemann presents absorption maximum at 570 nm for glyphosate when it is reacted with ninhydrin and sodium molybdate. Figure 2 presents the reaction scheme proposed for the colorimetric method of Ruhemann.

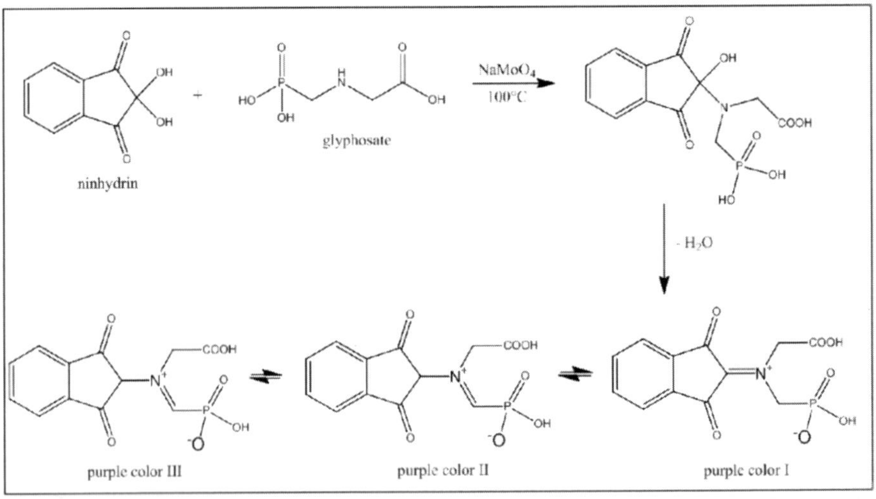

Figure 2. Reaction proposed to obtain the adduct in the colorimetric method of Ruhemann.

The quantification by HPLC was performed for glyphosate by submitting it to a derivatization reaction with an organic 9-fluorenylmethyl chloroformate compound (FMOC-Cl), using the method adapted from the literature (Catrinck et al., 2014). The scheme shown in Figure 3 represents the derivatization reaction of the glyphosate compound with FMOC-Cl.

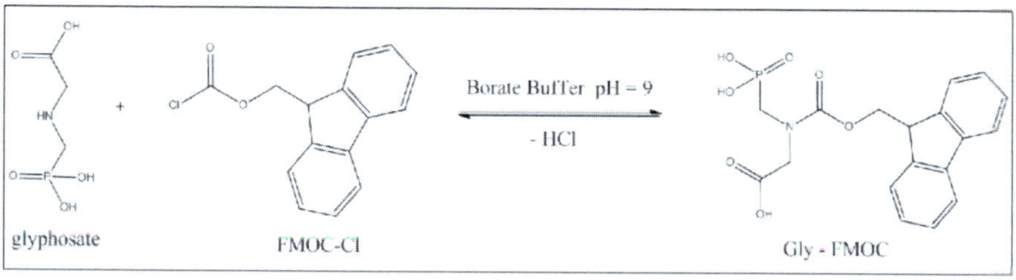

Figure 3. Derivatization reaction of glyphosate with FMOC-Cl.

These analyses were carried out in a C18 reverse-phase column (150 mm x 4.6 mm x 5 µm, Restek), with isocratic elution mode, with acetonitrile as the mobile phase (80%) and phosphate buffer prepared at pH 2.4 (20%), flow of 1.3 $mL.min^{-1}$ and detection at 260 nm. The peak of the derivatized glyphosate presented a retention time of 1.2 minutes, while the time for the FMOC-Cl reagent was 2.5 minutes.

RESULTS AND DISCUSSION

Infrared Spectroscopy

In a preliminary analysis of the data obtained for the glyphosate compound, it was possible to identify absorption bands characteristic of the organic functional groups present in the molecule. The complete spectrum of glyphosate in the solid state in the mid-infrared region is depicted in Figure 4. All the bands related to the absorption of this compound are listed below, accompanied by a discussion of the main bands that characterize the molecule.

Infrared spectrum of glyphosate (cm^{-1}, w = weak, m = medium, S = strong, sh = shoulder): 3012 (w), 3001 (w), 2970 (w), 2798 (sh), 2361 (m), 2341 (m), 1880 (w), 1730 (m), 1713 (m), 1554 (m), 1483 (m), 1468 (m), 1430 (m), 1421 (m), 1335 (w), 1269 (m), 1242 (m), 1220 (m), 1200 (m), 1149 (S), 1087 (S), 1076 (S), 1030 (m), 1000 (m), 980 (m), 910 (S), 884 (m), 829 (m), 797 (S), 775 (S), 669 (sh), 648 (m), 578 (m), 494 (S), 471 (S).

For better understanding, the vibrational spectrum of glyphosate presented above can be divided into three regions, based on the groups that have the possibility of interacting with the LDH.

The region in the spectrum between 1300 and 1800 cm^{-1} in general includes the bands associated with the vibrational modes of the carboxylate group present in the structure of glyphosate. The region located between 800 and 1200 cm^{-1}, in turn, mainly contains absorption bands related to the phosphonate group. Finally, the region between 3000 and 3500 cm^{-1} is typical of absorption modes belonging to the O-H bonds present in the molecule, possibly intensified by the presence of water.

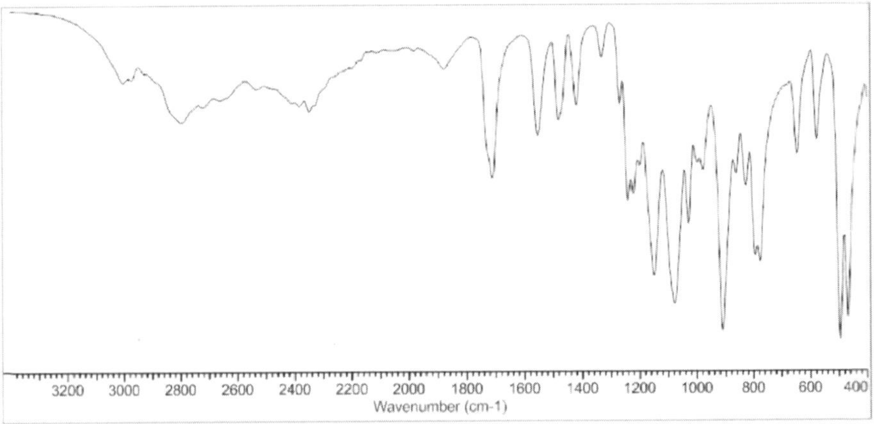

Figure 4. Infrared spectrum of the glyphosate compound in the solid state.

The ability to form hydrogen bonds, both intermolecular and intramolecular, due to their zwitterionic behavior and also the pH of the medium containing the glyphosate, influences the extension of those bonds and can affect the vibrational spectrum of the glyphosate molecule. According to Piccolo and Celano (1993), the stretching of the C=O bond present in the carboxylate group appears as two bands at 1730 and 1424 cm^{-1}, which in turn are shifted to 1720 and 1434 cm^{-1} when that group is involved in hydrogen bonds. The spectrum obtained in this work contained bands with medium intensity located at 1713 and 1430 cm^{-1}, indicating participation of the carboxylate group in the hydrogen bond interactions with neighboring molecules (Piccolo & Celano, 1994). Based on the works of Shoval (1981) and Subramaniam (1988), it was possible to attribute the other bands as being related to the carboxylate group. The spectrum confirmed the presence of the COO^- portion, through the bands with medium intensity corresponding to the asymmetric and symmetric stretching modes, present respectively at 1730 and 1421 cm^{-1} (Piccolo & Celano, 1994).

The bands located at 1174 and 1094 cm^{-1} were attributed, respectively, to the asymmetric and symmetric stretching of the P-O bonds of the phosphonate group due to the resonance between the P-O and P-OH forms. However, that attribution is not totally correct (Piccolo & Celano, 1994). According to the experiments performed by Piccolo & Celano (1993) about the interaction of glyphosate with humic acids, the intensity of that band is directly associated with the degree of dissociation of the group, with the total disappearance of this band at pH near 8. The symmetric stretching of the phosphonate group can be noted in the spectrum as a single band with strong intensity at 910 cm^{-1}, while the asymmetric stretching is decomposed into two bands, one observed at 1076 cm^{-1} and the other at 1149 cm^{-1}. The elongation of the P=O bond, in turn, is present in the band with absorption value of 1269 cm^{-1}. The amino group is present in the protonated form NH_2^+, confirmed by the bands at 3012 and 796 cm^{-1}, complemented by the presence of two other bands, one at 1554 and the other at 1483 cm^{-1}, corresponding to the angular deformation of the protonated amino group (Piccolo & Celano, 1994).

Just as for characterization of the organic compound glyphosate, infrared spectroscopy sheds light on the structural behavior of LDHs. For better understanding, the spectrum of the LDH in the Mg-Al-glyphosate system presented in Figure 5 is also subdivided into three fundamental regions. The first is from 3500 to 3000 cm^{-1}, where it is possible to identify the absorptions related to the OH present in the structure as a whole (OH from intercalated water and OH from the lamellar structure). The region from 1700 to 1400 cm^{-1}, in turn, is related to the vibrational modes present in the structure of the intercalated glyphosate anions and also the water molecules present. The region below 1400 cm^{-1} includes the vibrational modes correlated to the phosphonate group present in the glyphosate structure and also the vibrational modes of the O-M-O of the lamellar structure.

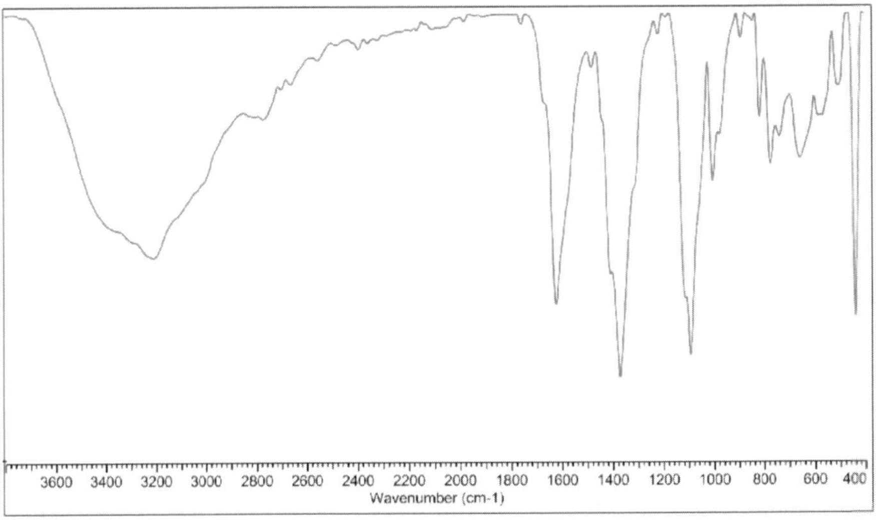

Figure 5. Infrared spectrum of the LDH in the Mg-Al-Gly hybrid system.

With respect to the OH groups belonging to the lamellar structure and the water molecules in the structure, various published studies have reported the presence of a band that covers a wide spectrum range, generally from 3400 to 3000 cm^{-1} (Halikia et al., 1998). This band might be the result of the overlap of two or even three distinct vibrations of the OH bonds of the intercalated water. Lutz (1995) proposed that the brucite-type hydroxides present a relatively intense band associated with the combination of vibrational modes in chain throughout the lamellar system, also called librational modes or network modes.

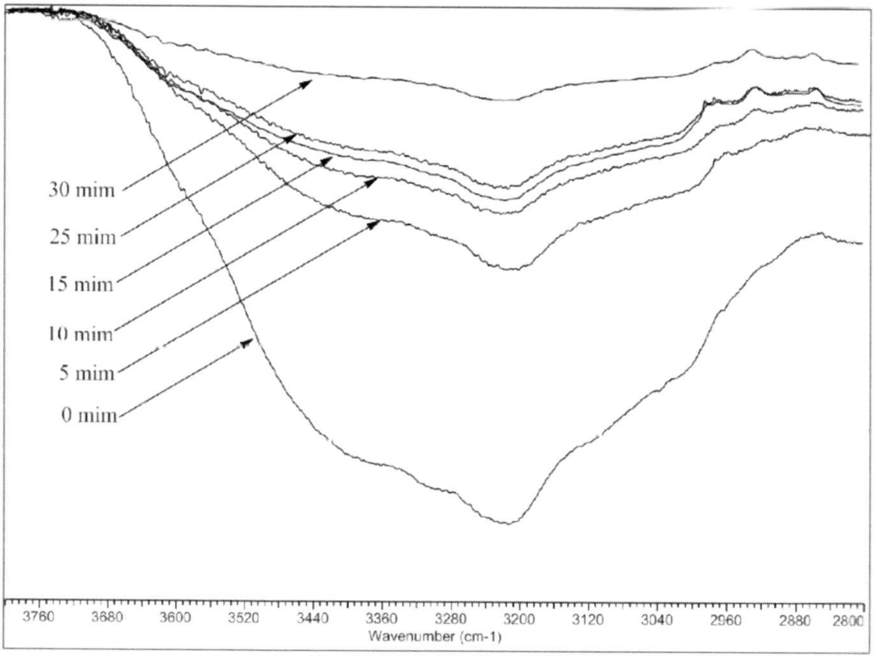

Figure 6. Infrared spectra in the ν_{OH} region measured after heating the LDH-glyphosate hybrid system at 250°C for time intervals of 5 to 30 minutes. The bands are arranged from the bottom to top, as indicated by the arrows.

Experiments involving heating samples of LDH containing glyphosate to 250°C and collecting their FTIR spectrum at intervals of 5 to 30 minutes showed strong dependence of the band in the region from 3500 to 3000 cm^{-1} on the vibrational modes related to the OH groups of the intercalated water molecules and/or those adsorbed on the surface. Analysis of the spectra in Figure 6 reveals that the area under the broad band tended to decline as the heating time increased, indicating that fewer OH groups of the water molecules contributed to that adsorption mode. This experimental evidence agrees with information found in the literature (Kloprogge & Frost, 2001).

Li and colleagues (2004) reported that the absorption maximums of the OH groups of the water molecules intercalated in LDH and also on its surface can present some variations within the region from 3500 to 3000 cm^{-1} due to the hydrogen bond interactions to which these groups are subject. Those interactions can occur both with the layer itself and also with the anions that are divided between the interlamellar domains. According to Li et al. (2004) and Li et al. (2005), these values tend to be very distinct in the case of the glyphosate present in the medium, since these molecules have various functional groups able to form hydrogen bonds.

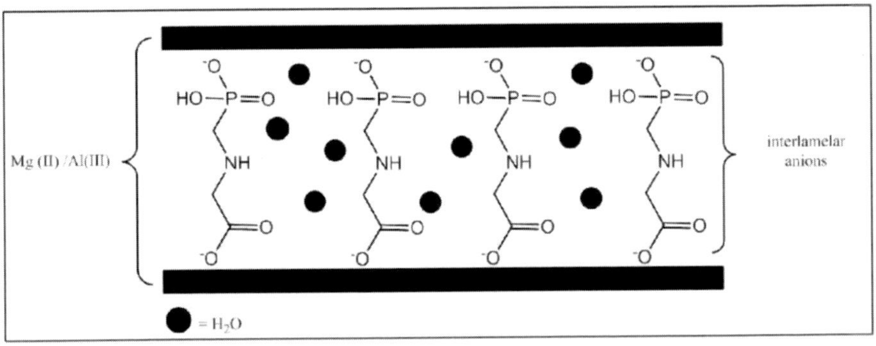

Figure 7. Structure proposed for the LDH in the Mg-AL-Gly hybrid system at pH 9.

The broad bands with absorption maximum at about 1620 cm^{-1} in the spectra of the materials obtained can be attributed to the asymmetric stretching of the COO^- groups present in the glyphosate molecules when they are interacting with the lamellae of the LDH. The presence of these groups can be confirmed by the presence of another broad band with absorption maximum near 1370 cm^{-1}, corresponding to the symmetric stretching mode of these same groups. Li (2005) and Meng (2005) attributed the broadening of the band with maximum at 1620 cm^{-1} to the absorption related to the flexion mode of the water and also the absorptions related to the nitrate present in the interlamellar space.

The phosphonate groups, when interacting with the LDH, in turn can be properly characterized by two broad bands, one with greater intensity than the other, having absorption maximums at 1100 and 980 cm^{-1}. Those bands, when not associated, were present at 1174 and 1094 cm^{-1}, respectively. Again due to the band overlap, Meng (2005) suggested that the absorption modes for the $P\text{-}O^{3-}$ in this region should be around 1058 and 1097 cm^{-1}.

Finally, the region between 800 and 400 cm^{-1} is characteristic of the vibrational modes referring to the M-O-M bonds and the harmonic vibrational mode of the structure as a whole (Li et al., 2005). The data from the literature on this region suggest that the vibrational modes between 900 and 600 cm^{-1} are related to the M-O bonds, while the O-M-O bonds are closer to 400 cm^{-1} (Li et al., 2004).

The data obtained by analyzing the infrared spectra together with the XRD diffractogram of the glyphosate compound and also the material of the Mg-Al-Gly system (pH 9) indicate the anions interact with the lamella through the carboxylate and phosphonate regions, as indicated by the structure presented in Figure 7 (Meng et al., 2005).

Solid-State Nuclear Magnetic Resonance Spectroscopy

The only stable aluminum isotope is ^{27}Al, which has a quadrupolar nucleus with a quantum nuclear spin number of I = 5/2. Its natural abundance of approximately 100%, which associated with a resonance frequency near that of ^{13}C results in good receptivity, making the NMR spectra of ^{27}Al relatively easy to obtain (Bräuniger et al., 2013). NMR is a very useful tool for structural characterization of compounds that contain aluminum nuclei, since the chemical signal of ^{27}Al is highly sensitive to the local coordination of the aluminum species that are present (Grey & Vega, 1995).

The NMR spectrum of ^{27}Al was obtained by an experiment with rotation of 8 kHz at room temperature using a 4 mm rotor, which enabled observing a signal referring to the main transition (+1/2/-1/2), as presented in Figure 8, with an isotropic chemical shift of 6.51 ppm, typical of hexacoordinated aluminum.

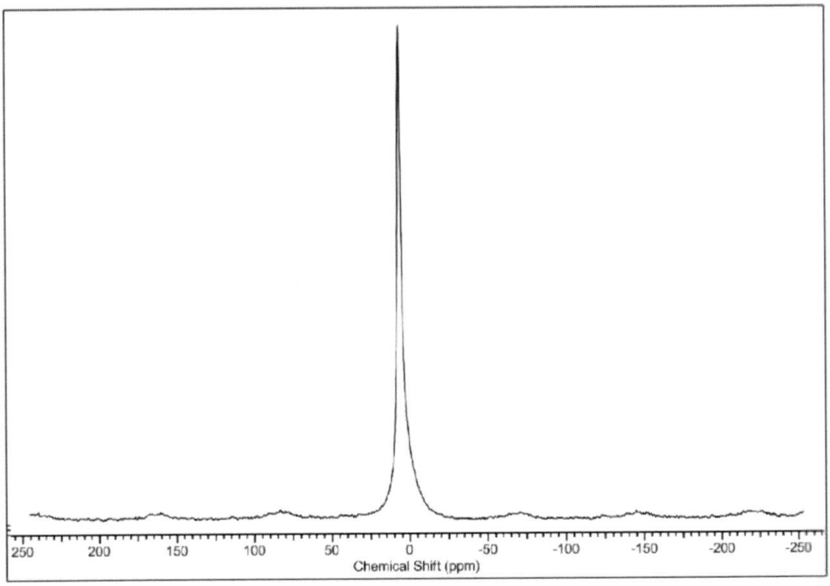

Figure 8. ^{27}Al CP-MAS NMR spectrum of the LDH-glyphosate compound.

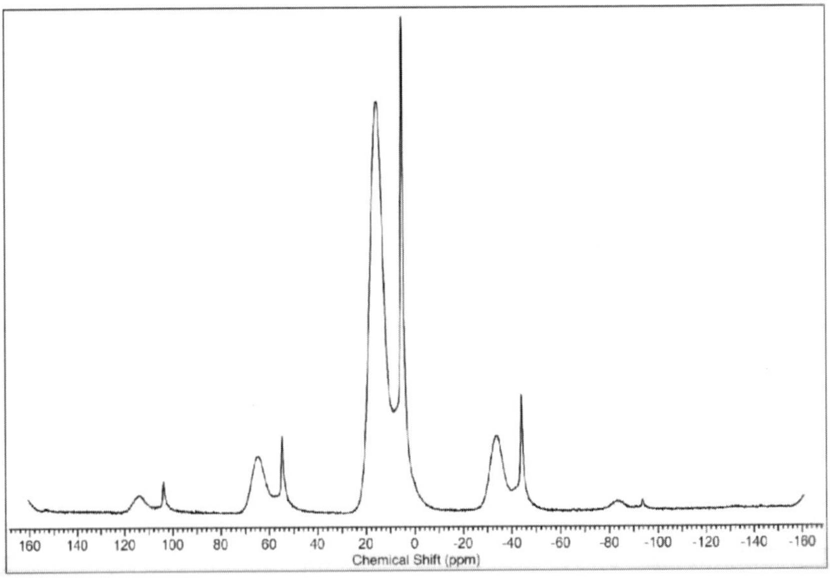

Figure 9. ^{31}P CP-MAS NMR spectrum of LDH-glyphosate compound.

The signal present in the spectrum with cubic symmetry indicates an isotropic charge distribution, i.e., all the ligands associated with the aluminum nucleus are equivalent (Roelofs et al., 2002; Rocha, 2001; Fyfe et al.; 2001).

Just like the ^{27}Al nucleus, the ^{31}P isotope, whose quantum nuclear spin number is I = 1/2, has natural abundance of practically 100% and 390-fold greater sensitivity to NMR than observed for the ^{13}C nucleus, resulting in excellent receptivity, making the experiments very practical. The chemical shifts observed for the ^{31}P nucleus cover a region from 500 to -200 ppm. This ample range demonstrates the strong sensitivity of the isotropic chemical shift of ^{31}P to the nature of the phosphorus atoms and their environments. The spectrum depicted in Figure 9 was obtained by a non-static experiment, and just as for aluminum, at rotation of 8 kHz.

The signals observed in the experimental condition used show axial symmetry and indicate, as mentioned, the chemical nature of the phosphorus nucleus. The spectrum obtained at 8 kHz was used only for informative purpose referring to the ^{31}P nuclei, since the signals in question did not have isotropic character (Laws et al., 2002). The narrower signal indicates the presence of non-intercalated phosphorus, possibly adsorbed on the surface. In turn, the broader signal, resulting from the weak dipole interactions, was associated with the phosphorus nuclei present in the interlamellar space (Greenwood et al., 2017; Rocha, 2001; Fyfe et al., 2001).

Powder X-Ray Diffraction of the LDH-Gly Compound

The diffractogram of the LDH-Gly compound presented in Figure 10 shows sharp and symmetric peaks of the planes (003), (006) and (110), as well as broad peaks of the planes (009), (015) and (018). These peaks are characteristic of a material with regular crystallinity and 3R1 rhombohedral symmetry (Meng et al., 2005).

By applying the 2θ value of the peak (003) and Bragg's equation, we obtained an interlamellar space of 0.69 nm. Further according to Meng et al. (2005), the minimum spacing to allow the presence of glyphosate in the interlamellar region would be approximately 0.40 nm, corresponding to the glyphosate linked horizontally to the lamellae. For values near 0.70 nm, it can be inferred that the glyphosate present in the interlamellar region is in a position near vertical in relation to the layers, as presented in Figure 7.

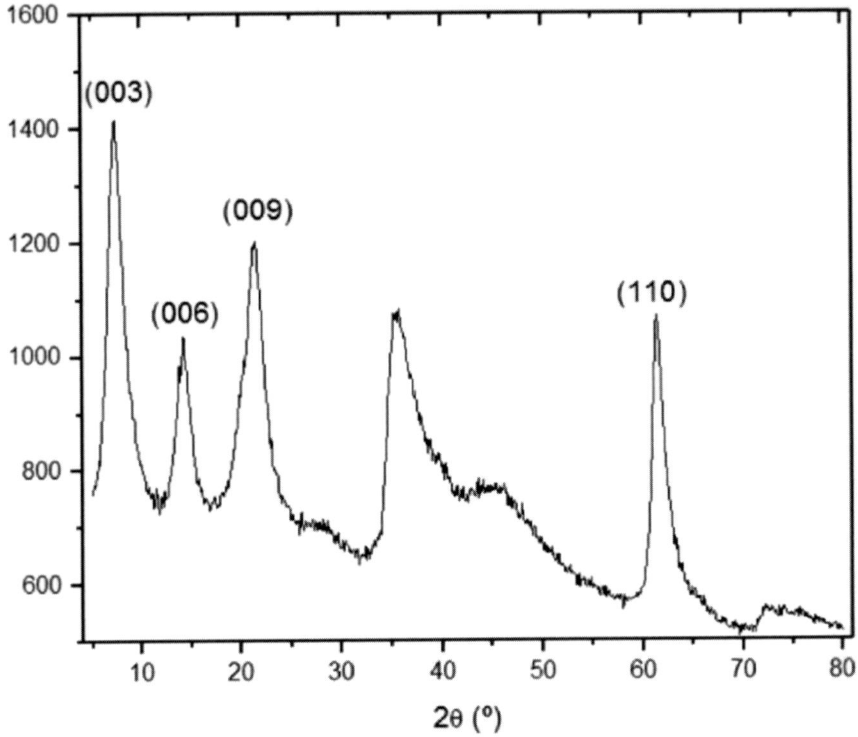

Figure 10. X-Ray powder diffractogram of the LDH-glyphosate compound.

Release Kinetics of Glyphosate from the LDH-Gly System

The release kinetics was studied in solutions containing carbonate, nitrate and bicarbonate anions in the following concentrations (mmol): 0.1, 0.4, 0.8, 1.5 and 2.5. The quantity of glyphosate released from each of them was analyzed by the colorimetric method of Ruhemann and also by HPLC.

The release of glyphosate from the LDH matrix can be attributed to the ion exchange process with the anions present in solution. The carbonate, nitrate and bicarbonate anions each present a specific degree of affinity for the double hydroxide layer. This hypothesis was confirmed by bin Hussein (2005) in a dynamic release study of the herbicide 2,4-D from a LDH matrix in the Zn-Al system. In that study, it was verified that the mass of 2,4-D present in the medium was proportional to the mass of LDH containing the herbicide used in the experiment, and that its release could be attributed to the possible ion exchange between the anions and carbonate coming from the atmospheric carbon dioxide dissolved in solution. Meng (2005) studied the release dynamics of glyphosate in a solution containing carbonate anions and verified the efficiency of the ion exchange during a period of time. They concluded that due to the strong affinity of the inorganic anions with LDH, immediate exchange occurred with the intercalated anions nearer the surface, causing a rapid increase in the release process, which soon tended to linear behavior due to the stabilization of the LDH by the carbonate, which prevented the exchange of the glyphosate anions located more internally.

The results of the release kinetics study in this work are reported in Table 1.

The data obtained in the release/ion exchange tests indicated that low concentrations of the inorganic anions reflected a minimal exchange process between the intercalated anions and those present in solution. This can be explained by the low glyphosate concentration in the medium. In turn, a possible explanation for that phenomenon is the stabilization property of LDH, since glyphosate has a -2 anionic charge, meaning two molecules of nitrate or bicarbonate are necessary to offset the charge of the organic anion. At higher concentrations, the exchange process is more effective, due to the greater availability of anions to offset the charge of glyphosate. The carbonate anions presented greater exchange efficiency than the other anions, because they have -2 anionic charge, equal to that of glyphosate (Li et al., 2005; Meng et al., 2005).

Table 1. Values obtained from the release studies

mmol of the anions	Concentration of glyphosate released ($mg.mL^{-1}$)		
	carbonate	nitrate	bicarbonate
0.1	1.14 (0.51)	1.10 (0.42)	1.12 (0.43)
0.4	2.04 (0.64)	1.21 (0.27)	1.53 (0.37)
0.8	2.77 (0.75)	1.48 (0.18)	1.88 (0.56)
1.5	2.92 (0.79)	1.66 (0.30)	1.77 (0.73)
2.5	2.96 (0.96)	1.66 (0.35)	2.24 (0.76)

The figures outside the parenthesis refer to data from colorimetric method of Ruhemann, and the figures in parenthesis refer to data from HPLC.

The results of this study indicate that the use of the material in cropped soil will be directly influenced by the physical and chemical factors of the medium. It is known that the ideal pH for cultivating many species is in a very restricted range, varying from 5.5 to 7. For values below this interval, the concentration of free metals in the soil increases substantially, which can impair the utility of the material, since the glyphosate released could form complexes with these metals, making its uptake by the plants impossible. For higher pH values, the high concentration of anions can result in a rapid release process, causing the glyphosate to accumulate freely in the soil, which is also not desired (Coutinho & Mazo, 2005; see also Agriculture and Food, 2017, for a scheme).

The results obtained by the glyphosate dosing techniques employed in this work allowed analyzing two distinct perspectives, one qualitative and the other quantitative. From the qualitative standpoint, both techniques converged to the same type of response regarding the greater quantity of glyphosate released from the solutions containing higher concentrations of inorganic anions. Hence, for a rapid evaluation and dynamic analysis of the glyphosate release process, the method of Ruhemann can be a more practical alternative. However, from the quantitative point of view, we observed that the method of Ruhemann overestimates the concentration values of the species in the medium compared to the data obtained by HPLC, so chromatography is more precise than that method.

Both analytic procedures are influenced by the effect of the matrix, but the derivatization method using FMOC-Cl eliminates this effect. Since FMOC-Cl is selective for glyphosate, once derivatized, the adduct formed will present greater sensitivity to chromatography, also allowing detection at distinct wavelengths of possible interferents in the medium, such as the residual nitrate that absorbs at 205 nm together with the non-derivatized glyphosate. We also observed that the presence of water in the reaction medium impaired the derivatization process of glyphosate, interfering in the symmetry of the peaks obtained in the chromatograms, and consequently the calculations to obtain the concentration of this anion in the medium.

CONCLUSION

Based in the findings of this work, we can conclude that the LDH Mg-Al-glyphosate prepared is a promising material for slow release of glyphosate in cropping systems. The procedure to synthesize the material was practical, indicating the possibility of scale-up production.

The infrared spectrum of the LDH Mg-Al-glyphosate presented bands attributed both to the host LDH and the glyphosate intercalated between the lamellae, and revealed that the material has regular crystallinity. The formation of the compound could be confirmed by the shifts attributed to the carboxylate groups (from 1730 to 1620 cm^{-1}) and phosphonate groups (region between 1170 and 910 cm^{-1}) of the glyphosate molecule. The variations in the bands in these regions suggested the interaction of both ends of the molecule with the lamellar system. This was subsequently confirmed by the presence of the 003 peak in the X-ray diffractogram used to obtain information about the basal spacing of the compound, were a value close to 0.7 nm was obtained, in accordance with intercalated glyphosate with vertical orientation. The test involving heating the material to 250°C showed that the vibrations related to the O-H bonds of the water molecules present in the material contributed intensely to the broad band present between 3000 and 3500 cm^{-1}, because the progressive elimination of water due to the heating also reduced the intensity of this band. The peak at 6.51 ppm in the ^{27}Al NMR spectrum with cubic symmetry helped confirm the integrity of the lamellar layer, since in this each nucleus should be associated with six identical ligands. The intercalation of the glyphosate anions was also confirmed by the ^{31}P NMR measurements in the solid state, which presented two distinct signals, one of them assigned to the intercalated glyphosate molecules.

By means of the release kinetics/ion exchange assays, we observed that the carbonate anions are more efficient in the exchange process, as indicated by the higher concentration of glyphosate present in the solution containing these anions. We also observed that due to the smaller charge associated with the nitrate and bicarbonate anions, they were unable to shift the glyphosate anions with the same efficiency as the carbonate, as revealed by the lower concentrations of glyphosate present in the solutions of these anions.

With respect to the techniques to quantify glyphosate, the high-performance liquid chromatography method presented greater precision and certainty. However, the method of Ruhemann, even though not having the same precision as HPLC, was efficient as a qualitative method, since it presented the same tendency regarding the effect of rising concentrations of glyphosate in solution. Therefore, the method of Ruhemann is a useful and inexpensive analytic tool for routine qualitative or semi-quantitative analyses in the laboratory to verify reactions involving glyphosate.

Concerning the derivatization procedure, this was only possible to achieve with reasonable efficiency when the reaction was conducted in a medium containing the lowest possible quantity of water, unlike reported previously in the literature. We observed that due to the low solubility of FMOC-Cl in the aqueous medium, it precipitated quickly during the derivatization process. The presence of water also interfered in the balance of the glyphosate and its zwitterionic form, also affecting the derivatization process. Another important aspect is the presence of borate in the medium. The derivatization process should be carried out at alkaline pH, maintained by a borate buffer, but these ions should be present in balanced form in the reaction medium, because despite being fundamental in the process of deprotonation of the amine group of glyphosate, if present in exaggerated quantities they promote the formation of precipitate, which impedes the analysis via HPLC.

The release dynamics study of glyphosate in a homogeneous medium can help shed light on the kinetics of this anion in heterogeneous media such as the soil, since anions such as carbonate, nitrate and bicarbonate will be present in their various forms, depending on the pH, and also in variable concentrations. It is possible to predict the release behavior, for example, in situations such as the process of soil correction with calcium carbonate for determined crops.

Finally, we can conclude that the use of hybrid compounds such as LDH Mg-Al-glyphosate can provide an environmental gain, by better utilization of the chemical properties of glyphosate, since it will be less susceptible to the negative effects of the weather.

REFERENCES

Acquavella, J. F.; Alexander, B. H.; Mandel, J. S.; Gustin, C.; Baker, B., Chapman, P.; Bleeke, M. *Environ Health Perspect*. 2004, *112*, 321 – 326.

Agriculture and Food (2017) https://www.agric.wa.gov.au/soil-acidity/effects-soil-acidity?page=0%2C1. Access 04/20/2018

Araújo, A. S.; Monteiro, R. T.; Abarkeli, R. B. *Chemosphere*. 2003, *52*, 799 – 804.

Arregui, M. C.; Lenardón, A.; Sanchez, D.; Maitre, M. I.; Scotta, R.; Enrique, S. *Pest Manag Sci*. 2004, *60*, 163 – 166.

Ashton, F. M.; Crafts, A. S. *Mode of Action of Herbicides*; 2nd Ed.; John Wiley and Sons: New York, NY, 1981; pp 236-253.

Barja, B. C.; Afonso, M. S. *Environ Sci Technol*. 2005, *39*, 585 – 592.

Battaglin, W. A.; Kolpin, D. W.; Scribner, E. A.; Kuivila, K. M.; Sandstrom, M. W. *J Am Water Resour Assoc*. 2005, *41*, 323 – 332.

Bhaskara, B. L.; Nagaraja, P. *Helv Chim Acta*. 2006, *89*, 2686 – 2693.

bin Hussein, M. Z.; Yahaya, A. H.; Zainal, Z.; Kian, L. H. *Sci Technol Adv Mater*. 2005, *6*, 956 – 962.

Bräuniger, T.; Jansen, M. *Z Anorg Allg Chem*. 2013, *639*, 857 – 879.

Brecke, B. J.; Duke, W. B. *Plant Physiol*. 1980, *66*, 656 – 659.

Campbell, W. F.; Evans, J. O.; Reed, S. C. *Weed Sci*. 1976, *24*, 22 – 25.

Carrado, K. A.; Kostapapas, A.; Suib, S. L. *Solid State Ionics*. 1988, *26*, 77 – 86.

Catrinck, T. C. P. G.; Dias, A.; Aguiar, M. C. S.; Silvério, F. O.; Fidêncio, P. H.; Pinho, G. P. *J Braz Chem Soc*. 2014, *25*, 1194 – 1199.

Constantino, V. R. L.; Pinnavaia, T. J. *Inorg Chem*. 1995, *34*, 883 – 892.

Constantino, V. R. L.; Pinnavaia, T. J. *Catal Lett*. 1994, *23*, 361 – 367.

Coutinho, C. F. B.; Mazo, L. H. *Quím Nova*. 2005, *28*, 1038 – 1045.

Crepaldi, E. L.; Pavan, P. C.; Tronto, J.; Valim, J. B. *J Colloid Interface Sci.* 2002, *248*, 429 - 442.

Crepaldi, E. L.; Pavan, P. C.; Valim, J. B. *J Mater Chem*. 2000, *10*, 1337 – 1343.

Crepaldi, E. L.; Valim, J. B. *Quím Nova*. 1998, *21*, 300 – 311.

Dick, R. E.; Quinn, J. P. *Appl Microbiol Biotechnol*. 1995, *43*, 545 – 550.

Duke, S. O.; Powles, S. B. *Pest Manag Sci*. 2008, *64*, 319 – 325.

Franz J. E.; In *The Herbicide Glyphosate;* Franz J. E.; Grossbard E.; Atkinson D.; Ed.; Butterworths: London, 1985; pp 3-17.

Fyfe, C. A.; Bretherton, J. L.; Lam, Y. L. *J Am Chem Soc*. 2001, *123*, 5285 – 5291.

Giesy, J. P.; Dobson, S.; Solomon, K. R. *Rev Environ Contam Toxicol*. 2000, *167*, 35 – 120.

Glass, R. L. *J Agric Food Chem*. 1987, *35*, 497 – 500.

Gomez, M. A.; Sagardoy, M. A. *Rev Latinoam Microbiol*. 1985, *27*, 351 – 357.

Greenwood, A. I.; Clay, M. C.; Rienstra, C. M. *J Magn Reson*. 2017, *278*, 8 – 17.

Grey, C. P.; Vega, A. J. *J Am Chem Soc*. 1995, *117*, 8232 – 8242.

Haderlie, L. C.; Slife, F. W.; Butler, H. S. *Weed Res*. 1978, *18*, 269 – 273.

Halikia, I.; Neou-Syngouna, P.; Kolista, D. *Thermochim Acta*. 1998, *320*, 75 – 88.

Hensley, D. L.; Beuerman, D. S. N.; Carpenter, P. L. *Weed Res*. 1978, *18*, 287 – 291.

Hoagland, R. E.; Duke, S. O.; Elmore, C. D. *Physiol Plant*. 1979, *46*, 357 – 366.

Jaworski, E. G. *J Agric Food Chem*. 1972, *20*, 1195 – 1198.

Kloprogge, J. T., Frost, R. L. In *Layered Double hydroxides: Present and Future*; Rives, V.; Ed.; Nova Science Publishers: New York, NY, 2001; pp 153 – 215.

Kostadinova, D.; Pereira., A. C.; Lansalot, M.; D'Agosto, F.; Bourgeat-Lami, E.; Leroux, F.; Taviot-Guého, C.; Cadars, S.; Prevot, V. *Beilstein J Nanotechnol*. 2016, 7, 2000 – 2012.

Lane, M.; Lorenz, N.; Saxena, J.; Ramsier, C.; Dick, R. P. *Pedobiologia*. 2012, *55*, 335 – 342.

Laws, D. D.; Bitter, M. H.; Jerschow, A. *Angew Chem Int Ed Engl*. 2002, *41*, 3096 – 3129.

Li, F.; Zhang, L.; Evans, D. G.; Forano, C.; Duan, X. *Thermochim Acta*. 2004, *424*, 15 – 23.

Li, F.; Wang, Y.; Yang, Q.; Evans, D. G.; Forano, C.; Duan, X. *J Hazard Mater. B*. 2005, *125*, 89 – 95.

Liu, B.; Dong, L.; Yu, Q.; Li, X.; Wu, F.; Tan, Z.; Luo, S. *J Phys Chem A*. 2016, *120*, 2132 – 2137.

Lutz, H. D., In: *Coordination Chemistry. Structure and Bonding*, Vol. 82. Springer: Berlin, Heidelberg, 1995; pp 85 – 103.

McConnell, J. S.; Hossner, L. R. *J Agric Food Chem*. 1985, *33*, 1075 – 1078.

Meng, J.; Zhang, H.; Evans, D. G.; Duan, X. *Chin Sci Bull*. 2005, *50*, 745 – 751.

Miles, C. J.; Moye, H. A. *J Agric Food Chem*. 1998, *36*, 486 – 491.

Miyata, S. *Clays Clay Miner*. 1983, *31*, 305 – 311.

Peixoto, M. M.; Bauerfeldt, G. F.; Herbst, M. H.; Pereira, M. S.; da Silva, C. O. *J Phys Chem A*. 2015, *119*, 5241 – 5249.

Peterson, C. A.; de Wildt, P. P. Q.; Edgington, L. V. *Pestic Biochem Physiol*. 1978, *8*, 1 – 9.

Piccolo, A., Celano, G. *Environ Toxicol Chem*. 1994, *13*, 1737 – 1741.

Piccolo, A.; Celano, G. *J. Environ Sci Health B*. 1993, *28*, 447 – 457.

Piccolo, A.; Celano, G.; Conte, P. *J Agric Food Chem*. 1996, *44*, 2442 – 2446.

Prata, F.; Lavorenti, A.; Regitano, J. B.; Vereecken, H.; Tornisielo, V. L.; Pelissari, A. *R Bras Ci Solo*. 2005, *29*, 61 – 69.

Prinetto, F., Ghiotti, G., Graffin, P., Tichit, D. *Microporous Mesoporous Mater*. 2000, *39*, 229 – 247.

Pushparaj, S. S. C.; Forano, C.; Prevot, V.; Lipton, A. S.; Rees, G. J.; Hanna, J. V.; Nielsen, U. G. *J Phys Chem C*. 2015, 119, 27695 – 27707.

Quinn, J. P., In *Pesticides interactions in crop production - beneficial and deleterious effects;* Altman, J.; Ed.; CTC Press: Boca Raton, FL, 1993; pp 245 – 265.

Reichle, W. T. *Solid State Ionics*. 1986, *22*, 135 – 141.

Rocha, J., In *Layered Double hydroxides: Present and Future*; Rives, V.; Ed.; Nova Science Publishers: New York, NY, 2001; pp 217 – 240.

Roelofs, C. A. A. J.; van Bokhoven, J. A.; van Dillen, A. J.; Geus, J. W.; de Jong, K. P. *Chem Eur J*. 2002, *8*, 5571 – 5579.

Rueppel, M. L.; Brightwell, B. B.; Schaefer, J.; Marvel, J. T. *J. Agric. Food Chem.* 1977, *25*, 517 – 528.

Scarpellini, D.; Leonardi, C.; Mattoccia, A.; Di Giamberardino, L.; Medaglia, P. G.; Mantini, G.; Gatta, F.; Giovine, E.; Foglietti, V.; Falconi, C.; Orsini, A.; Pizzoferrato, R. *J Nanomater*. 2015, Article ID 809486.

Shoval, S.; Yariv, S. *Agrochimica*. 1981, *25*, 377 – 386.

Subramaniam, V.; Hoggard, P. E. *J Agric Food Chem*. 1988, *36*, 1326 – 1329.

Theiss, F. L.; Ayoko, G. A.; Frost, R. L. *Appl Surf Sci*. 2016, *383*, 200 – 213.

Yamamoto, T.; Kodama, T.; Hasegawa, N.; Tsuji, M.; Tamaura, Y. *Energy Convers Manag*. 1995, *36*, 637 – 640.

Yang, W.; Kim, Y.; Liu, K. T. P.; Sahimi, M.; Tsotsis, T. T. *Chem Eng Sci*. 2002, *57*, 2945 – 2953.

Zablotowicz, R. M.; Reddy, K. N. *J Environ Qual*. 2004, *33*, 825 – 831.

In: Glyphosate
Editor: Ron Quinn

ISBN: 978-1-53614-002-6

Chapter 3

CONTRIBUTION TO THE STUDY OF GLYPHOSATE COMPLEXATION WITH NI (II)

Nazareth Ferreira da Fonseca and Marcelo Hawrylak Herbst*

Laboratório de Síntese Inorgânica e Materiais, Departamento de Química, Universidade Federal Rural do Rio de Janeiro, Seropédica, RJ, Brazil

ABSTRACT

The Ni (II) cation is essential for symbiotic N_2 fixation by higher plants, and there are conflicting reports in the literature concerning a possible correlation of the continuous use of glyphosate on crops with the decrease of Ni (II) bioavailability in the soil, impairing the growth of some crops, including genetically modified and glyphosate-resistant soybeans. On the other hand, Kutman et al. (2013) reported that foliar nickel application alleviates detrimental drift effects of glyphosate on wheat plants. However, the coordination and structural chemistry of Ni (II) – glyphosate complexes at different pH values and Ni (II): glyphosate stoichiometric ratios have been little studied. In this contribution, the

* Corresponding Author Email: herbst_marcelo@yahoo.com.br.

chelating capacity, coordination modes and structural chemistry of glyphosate with Ni (II) cations were investigated in solution and in the solid state. Glyphosate was purified from commercial Roundup WG (Monsanto), and characterized by ^{31}P NMR, FTIR and melting point. The nickel complexes were prepared from $Ni(NO_3)_2$ solutions at pH = 8.0 and 4.0, and characterized by solid-state infrared spectroscopy (4000-150 cm^{-1}) and elemental analysis (CHN). The solids were also studied by thermal analysis and X-ray absorption spectroscopy (XAS) using synchrotron radiation (LNLS, Brazil). The monitoring of the reactions in solution, through the analysis of the ligand field transition bands ($^3A_{2g} \rightarrow {}^3T_{2g}$) by electronic spectroscopy, revealed that complexation occurs in the experimental conditions employed. Infrared spectroscopy showed that in all cases the carboxylate group of glyphosate coordinates in a monodentate fashion, and coordination of the phosphonate and amino groups also occurs in basic media. In some cases, there was indication that water molecules participate in the coordination. It was possible to propose structures from the data for the products of the reactions between glyphosate and Ni (II) at pH = 8.0 and 4.0. At acid pH, the IR results, elemental and thermal analysis, and XAS spectra suggested that the complexes formed are the same regardless of stoichiometry. The results obtained from the simulation of the EXAFS spectra agreed with the vibrational spectroscopy analysis, in the sense that the Ni (II) complex obtained with 1:1 stoichiometry and acid pH was formed by only one coordination sphere composed of six light atoms, presumably oxygens, while the complex obtained at basic pH with 2:1 stoichiometry, on the other hand, was formed by two units, an anionic $[Ni(glyphosate)_2]^{2-}$ and a cationic $[Ni(H_2O)_6]^{2+}$.

Keywords: glyphosate, nickel, coordination chemistry, EXAFS

INTRODUCTION

In soils, nickel is a trace element that forms chelate complexes with organic compounds at the surface, and it generally presents higher stability constants than other metals. Nickel participates in the structure of the enzyme urease, and influences the hydrogenase enzymatic complex, enhancing the efficiency of symbiotic nitrogen fixation by soil microorganisms. It also participates in the synthesis of phytoalexines, which are useful for plant resistance against diseases (Malavolta, 2007).

There are conflicting reports in the literature concerning possible correlation of the continuous use of glyphosate in crops with the decrease of Ni (II) bioavailability in the soil, impairing the growth of some crops, including genetically modified and glyphosate resistant soybeans (Zobiole, 2010; Duke, 2012). On the other hand, Kutman et al. (2013) reported that foliar nickel application alleviates detrimental drift effects of glyphosate on wheat plants.

Concerning the coordination chemistry of glyphosate, as a multifunctional molecule it is a very interesting ligand. Thus the investigation of the modes of coordination of glyphosate in different conditions of pH and relative nickel concentration is important both from the point of view of structural inorganic chemistry and agricultural chemistry.

It is well known that the glyphosate molecule has three potential functional groups, and can act as a monodentate or polydentate ligand. The phosphonate group R-$PO(OH)_2$ is able to form stable complexes with metallic cations, both soluble and insoluble ones, and this ability strongly influences adsorption, photodegradation and biodegradation processes (Barja, 2001). Besides the phosphonate group, glyphosate also has two other functional groups, amino and carboxylate, which coordinate with metallic ions, especially the transition metal cations, near neutral pH. It was also observed that the adsorption of glyphosate in soil diminishes in basic pH, as a result of the repulsion of the negative charges of the anionic glyphosate and the residual negative charges of soil particles (Undabeytia, 2000).

Recently we studied all possible glyphosate conformations in aqueous solution that can be involved in deprotonation reactions in the pH range from 2 to 11, using the polarizable continuum method (PCM). We also compared direct (or absolute) methods to calculate pKa values, the cluster-continuum model and the proton-exchange scheme, using different thermodynamic cycles. The best result was achieved when using a proton-exchange scheme, which was able to properly reproduce three glyphosate experimental pKa values predicted for the glyphosate structures and conformations previously determined (Peixoto et al., 2015). Zwitterionic forms of the neutral species were found as stable structures by the PCM, albeit not abundant at room temperature. The most abundant zwitterionic structure with a formal charge of −1 lost protons from both the carboxylate and the phosphonate groups, but the amino group remained protonated. As a practical result, in the 7 - 8 pH range – typical phloem pH – the amino group should be non-protonated. Later, Liu et al. (2016) corroborated our findings experimentally, through a thermodynamic study using potentiometry, calorimetry and NMR spectroscopy.

When adsorbed onto soil particles, glyphosate tends to be inactive, supposedly due to bonding of the phosphonate group to ions, oxides and minerals, similar to inorganic phosphate. Indeed, the competition between glyphosate and inorganic phosphate was investigated *in vitro* (Undabeytia, 2000). Specific ions (Fe^{2+}, Fe^{3+}, Al^{3+}) coordinate to the glyphosate molecule and to humic acids present in the soil. This is important to understand the degradation mechanism of glyphosate when adsorbed in the soil. The free glyphosate is readily degraded through microbiotic activity, but the adsorbed glyphosate may be impossible to degrade, remaining inactive in the soil for years (Amarante and Santos, 2002).

Concerning the modes of coordination of glyphosate to the nickel (II) cations, there are only a few reports in the literature. Prenesti and Gulmini (1998) investigated the formation of complexes of Ni(II) and glyphosate in solution, but they were unable to propose any structure for the complexes. Nevertheless, the small values of ε_{max} measured for the complexes indicated octahedral geometry. Menelaou and colleagues (2009) described the synthesis, spectroscopic investigation, thermal stability and the crystal structure of a nickel-glyphosate complex. They obtained a single-crystal of a 2:1 (ligand to metal) complex in basic media. The geometry around the metallic center of Ni (II) is octahedral, and the glyphosate molecules act as tridentate ligands, producing only the facial isomer of an anionic complex, with hexaaquonickel (II) molecules as the counter ions.

Due to the scarcity of reports in the literature, here we present a contribution to the study of glyphosate complexation with Ni (II), so that future investigations can benefit from our findings.

MATERIALS AND METHODS

All solvents and reagents were used as received (Vetec/Sigma Aldrich, Brazil), with the exception of glyphosate, which was purified from Monsanto's Roundup WG and isolated as shiny crystals after treatment with strongly acidic solutions. The glyphosate in its acid form thus obtained was characterized by melting point (230 ± 1°C) and ^{31}P NMR (δ = 8.57 in D_2O, 161.9 MHz for the ^{31}P nucleus, measured in a 400 MHz Bruker Ultrashield spectrometer).

The syntheses of the complexes were carried out at room temperature, by dissolving 0.20 g (1.18 mmol) of glyphosate in the acid form in 20.0 mL of deionized water, followed by the addition of the appropriate amount of $Ni(NO_3)_2.6H_2O$, under magnetic stirring. Syntheses were performed at acidic and basic pH, and the nickel:glyphosate stoichiometry at each pH was 1:1 and 1:2.

The pH was adjusted by adding diluted NaOH solution, and the mixture was left under stirring for 12 hours to ensure completion of the reactions. Afterwards, the solvent was partially removed in a rotary evaporator at 60°C, and cold diethyl ether was gently added at room temperature to the concentrated solution, which was left to stand in a refrigerator for a few days, until precipitation occurred. Then, the supernatant was removed by filtration and the solids were dried in a desiccator under silica, with yield of around 85%. Despite our attempts to obtain single crystals of the complexes, all solids were obtained as fine water-soluble powders. With the exception of the 2:1 complex obtained at pH = 8, which was blue, all other complexes were green in color.

The complexes were characterized by FTIR spectroscopy (Nicolet Magna IR 560, CsI pellets, from 4000 to 150 cm^{-1}), electronic spectroscopy (Shimadzu UV-Mini, quartz cuvettes, in water solutions), elemental analysis (PerkinElmer), thermogravimetric analysis (Netzsch STA449 Jupter, TG/DSC sensor, 75 µL Al_2O_3 crucibles, 20mL/min He flux and 10 °C/min heating rate), and by EXAFS (XAFS-2 beamline at the Brazilian National Synchrotron Radiation Laboratory - LNLS using boron nitride pellets covered with Kapton tape, using a metallic nickel foil (15 μm) as E_0 standard). Data treatment of EXAFS spectra was performed by the Athena software package and the simulations were performed by using the FeFF6 freeware program.

Results and Discussion

FTIR Spectra

By comparing the solid-state FTIR spectra of nickel-glyphosate complexes with the spectrum of glyphosate, we obtained important information on the coordination modes of glyphosate. Table 1 lists the main absorption bands observed in the FTIR spectra of glyphosate, based on the assignments of Subramaniam (1988) and Jing (2018). The spectra of the nickel complexes of glyphosate were also measured in the 4000 – 150 cm^{-1} range, and are shown in Figure 1. Analysis of the FTIR spectra of the complexes was carried out based on Menelaou (2009) and Jing (2018), and the main absorptions are listed in Table 2.

Usually, the difference between the wavenumber values of the anti-symmetric and symmetric stretching of the carboxylate group ($\Delta = \nu_{as}\text{-}\nu_s$) is related to three coordination modes of the carboxylate: monodentate, bidentate (or chelate), and bridged (Nakamoto, 1997).

Table 1. Selected bands observed in the FTIR spectra of glyphosate in the solid state

Assignment	ν (cm^{-1})
ν (NH_2^+)	3015
ν COOH	1717, 1437
ν_{as} (COO^-)	1733
ν_s (COO^-)	-
δ (NH_2^+)	1554, 1481
ν_{as} (PO_2)	1166
ν_s (PO_2)	1085
ν (P-OH)	918

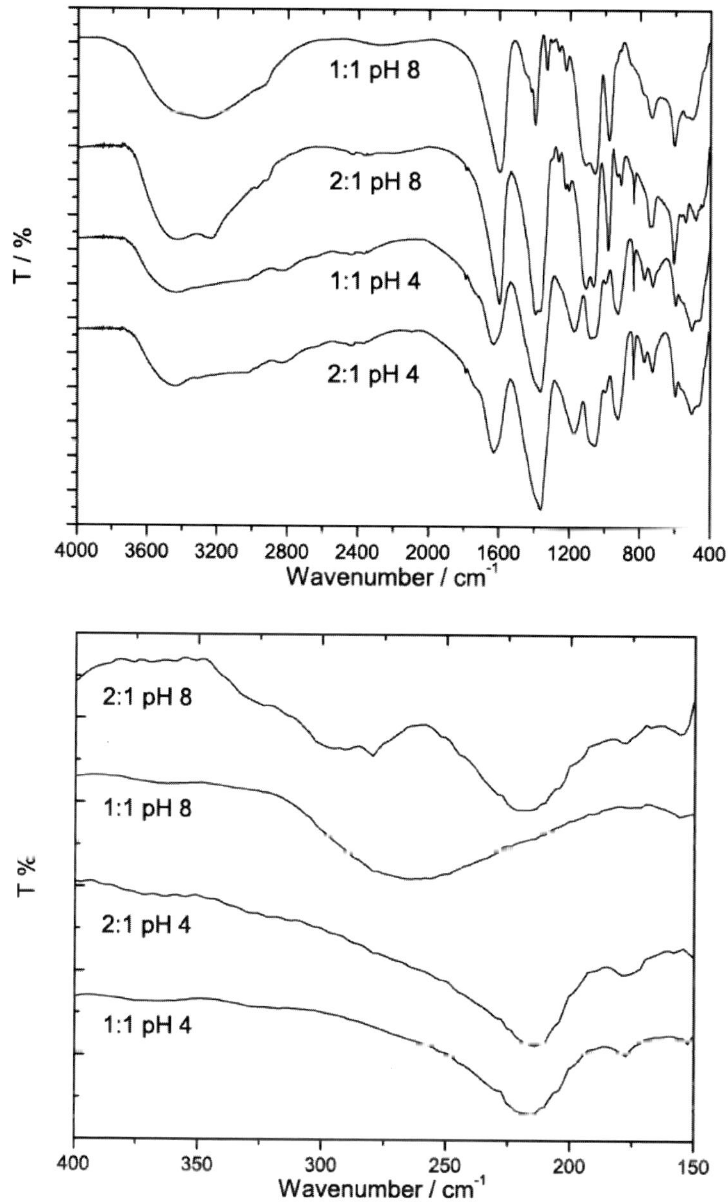

Figure 1. FTIR spectra of the nickel complexes of glyphosate in the 4000-400 cm^{-1} (top) and 400-150 cm^{-1} range (bottom).

Table 2. Selected bands observed in the FTIR spectra of nickel complexes of glyphosate

Main bands/cm^{-1}				
	2:1 pH = 8	1:1 pH = 8	2:1 pH = 4	1:1 pH = 4
$\nu(OH)H_2O$	3448	3414	3448	3449
$\nu(NH_2^+)$	3268	3175	3296	3269
ν_{as} COO^-	1597	1597	-	-
δ (NH)	-	-	-	-
ν_s COO^-	1428	1428	1428	1428
ν_{as} (PO_2)	1105	1107	1173	1174
ν_s (PO_2)	1064	1057	1057	1079
ν (P-OH)	980	976	926	926

Evidently, each of those coordination modes has different values for the wavenumbers of the antisymmetric and symmetric stretching vibrations of the carboxylate group. Thus, in complexes in which the carboxylate is bonded in a monodentate fashion, the Δ values are higher than those of the non-complexed carboxylate, but in complexes in which the carboxylate is bonded in a bidentate fashion, that value is considerably lower than for the non-complexed carboxylate. On the other hand, the bridged coordination of the carboxylate presents values of Δ higher than for the bidentate mode, but similar to the non-complexed carboxylate (Boyd, 1981).

We were not able to calculate the value of Δ for the carboxylate group of non-complexed glyphosate, since the carboxylate symmetric stretching band was superposed on bands assigned to other vibrational modes. Nevertheless, it was possible to calculate the Δ values for the nickel complexes, finding a value of 169 cm^{-1} for all four complexes. In this respect, according to Nakamoto (1997), that value is typical of monodentate carboxylate coordination.

Concerning the phosphonate group, changes were also observed in the region of the FTIR spectra that covers the vibrational modes of this group, suggesting complexation. The complexation induces a loss of resonance of the PO_3^{-2} group, as the oxygen atoms are bonded to the metal cations. As a consequence, the lower symmetry of the complexed phosphonate group leads to the splitting of the absorption bands (Undabeytia, 2002).

The low-frequency range of the FTIR spectra involves several coupled metal-ligand vibrational modes, and it was not possible to securely assign the bands to a specific vibration. However, in the 2:1 complex synthesized at pH = 8, it was possible to assign bands to both Ni-N and Ni-O vibrations, which occurred at 448 cm^{-1} and 280 cm^{-1}, respectively. A band observed at 405 cm^{-1} in the spectra of the complexes synthesized at pH = 4 was attributed to the coordination of water molecules.

Concerning the phosphonate $\nu_{M\text{-}O}$ vibration modes, it was not possible assign the bands due to the lack of reports in the literature. However, for the complexes synthesized at pH = 4, the FTIR spectra in the phosphonate $\nu_{M\text{-}O}$ range were the same, suggesting they are isostructural, regardless the initial stoichiometry.

Electronic Spectra

The divalent nickel cation has the electronic configuration [Ar]$3d^8$, and when in octahedral symmetry, it presents two unpaired electrons, regardless the magnitude of the ligand field splitting parameter, Δo. Thus, according to the Tanabe-Sugano diagram for the ^3F free ion spectral term, the fundamental electronic state is $^3A_{2g}$, and the spin-allowed transitions are those between the fundamental electronic state and the triplet excited electronic states, that is, $^3A_{2g} \rightarrow {}^3T_{2g}$, $^3A_{2g} \rightarrow {}^3T_{1g}(F)$ and $^3A_{2g} \rightarrow {}^3T_{1g}$ (P), where (F) and (P) indicate the electronic states that originate from the free ion spectral terms ^3F and ^3P, respectively. From the Tanabe-Sugano diagram, the $^3A_{2g} \rightarrow {}^3T_{2g}$ transition corresponds to the electronic transition of lowest energy, and is a measure of the Δo parameter. The electronic spectra of the nickel-glyphosate complexes in aqueous solutions were recorded at room temperature as a function of the solution pH, and are shown in Figure 2.

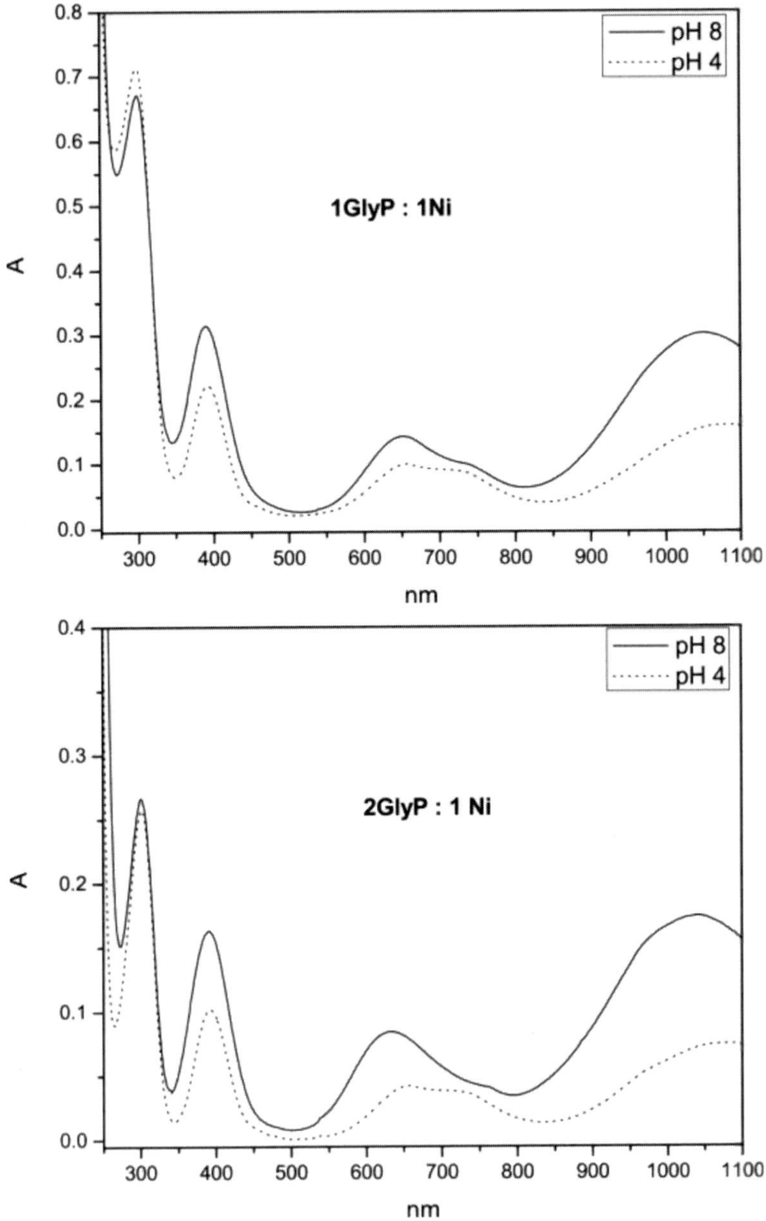

Figure 2. Electronic spectra of the Ni(II)-glyphosate complexes at different pH values. (top) 1 glyphosate: 1 Ni(II); (bottom) 2 glyphosate: 1 Ni(II).

Table 3. Ligand-field splitting parameters (Δo) measured for the Ni(II)-glyphosate complexes

Complex	$[Ni(H_2O)_6]^{2+}$ (*)	GlyP:Ni (pH 4)	2GlyP:Ni (pH 4)	GlyP:Ni (pH 8)	2GlyP:Ni (pH 8)	$[Ni(gly)_3]^-$ (**)
Δo (cm^{-1})	9000	9260	9300	9525	9615	10100

Source: (*) Cotton and Wilkinson, 1980; (**) Theriot et al. 1969, gly = glycinate.

The electronic spectra shown in Figure 2 indicate that in acidic pH the predominant species is the $[Ni(H_2O)_6]^{+2}$ complex, and that raising the pH value causes a considerable change in the Ni(II) coordination, followed by lower wavenumbers and consequently higher Δo values. Once the Δo parameter is obtained for different complexes, it can be used to construct a spectrochemical series of ligands, as shown in Table 3.

In aqueous solution the water molecules will compete with the glyphosate for the Ni(II) coordination, and the number of coordination sites occupied by the glyphosate ligands may be different compared with the solid-state species.

The results discussed above suggest that in acidic solution, more water molecules are coordinated to the Ni(II) metallic center, but in basic pH the water molecules are substituted by the glyphosate amino moiety.

Elemental Analysis

The results of elemental analysis of the nickel (II) – glyphosate complexes are in Table 4.

As shown, we were unable to propose any coherent formula for the complexes synthesized at pH = 4, but the results for the two initial stoichiometric ratios are quite close, reinforcing our conclusion from the FTIR spectra that at acidic pH a single compound is produced, regardless the initial stoichiometry.

On the other hand, the results suggest that at basic pH the different initial stoichiometry affords different complexes. The product obtained in equimolar ratio is consistent with the glyphosate coordinated in tetradentate fashion by the carboxylate oxygen, the amino nitrogen and two phosphonate oxygens. Completing the coordination, there are two water molecules. In contrast, the product obtained with excess of glyphosate is consistent with an anionic $[Ni(glyphosate)_2]^{2-}$ and a cationic $[Ni(H_2O)_6]^{2+}$ complex, in the same way as the complex obtained by Menelaou et al. (2009). Figure 3 shows the structures proposed based on the FTIR and elemental analyses.

Table 4. Elemental analysis (C, H, N) of the complexes

Experimental (Calcd.)/%			
Complex	C	H	N
2Glyphosate:1Ni(II) pH = 8	11.49 (12.92)	3.12 (3.98)	7.68 (6.02)
1Glyphosate:1Ni(II) pH = 8	12.58 (12.73)	2.65 (2.82)	4.54 (4.95)
2Glyphosate:1Ni(II) pH = 4	13.21 (-)	1.90 (-)	8.39 (-)
1Glyphosate:1Ni(II) pH = 4	12.93 (-)	2.67 (-)	8.59 (-)

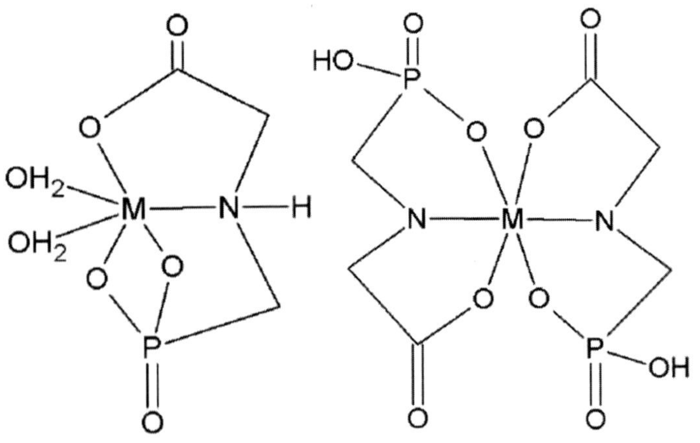

Figure 3. Proposed structures for the complexes obtained in basic pH. (left) 1GlyP: 1Ni(II); (right) 2GlyP: 1Ni(II).

The crystalline structure of a Cu(II)-glyphosate complex was reported by Clarke et al. (1989). The structure shows the metallic center coordinated to five donor atoms, in a distorted pyramidal geometry. Each Cu(II) cation is coordinated to an amino nitrogen, a carboxylate oxygen and a phosphonate oxygen of a single glyphosate molecule. Completing the coordination, there are two phosphonate oxygens of adjacent glyphosate molecules, rendering a polymeric structure.

Subramaniam and Hoggard (1988) obtained glyphosate complexes of Cu(II), Fe(II) and Ni(II), using a two-fold molar excess of glyphosate. They reported a green precipitate with low solubility in water, which was characterized by X-ray powder diffraction, elemental analysis and FTIR spectroscopy. They concluded that the glyphosate molecule can be tetra-coordinated to the metallic center, through amino nitrogen, carboxylate oxygen and two phosphonate oxygens.

Heineke et al. (1994) described the synthesis and crystalline structure of a Co(III) glyphosate complex. Due to the diamagnetic nature of the Co(III) center, it was possible to investigate the structure of the complex in solution by ^{1}H-NMR spectroscopy, finding it to be a complex mixture of isomers.

Han et al. (2008) reported coordination polymers of Co(II) and glyphosate, and were able to determine the crystalline structures. The structure of one of those coordination polymers was composed solely of phosphonate oxygens bonded to the Co(II) center, in a distorted octahedral geometry, whereas the structure of the second coordination polymer was composed of six oxygen atoms bonded to the Co(II) center, which could be grouped in three pairs of monodentate phosphonate, carboxylate and water ligands.

Thermogravimetric Analysis

To understand the structural features of the complexes, we investigated their thermal decomposition under inert He flux. For the complexes obtained at basic pH, there were some differences: below 300°C, the 2:1

complex showed two mass loss events, corresponding to 28.7% of the initial mass, but the 1:1 complex showed three losses, corresponding to 10.6% of the initial mass. These results are in accordance with the FTIR and elemental analysis discussed before.

Concerning the complexes obtained in acidic pH, there were few difference in the thermal stability below 300°C: the 2:1 complex showed three losses, corresponding to 21.14% of the initial mass, and the 1:1 complex showed the same three losses, corresponding to 23.3%. This result again corroborates our conclusion that at acidic pH, a single complex was obtained. Table 5 lists the temperatures and mass losses below 300°C for the complexes, all attributed to the loss of water molecules. Such attribution was based on Menelaou et al. (2009), who investigated by thermogravimetric analysis the crystalline $[Ni(H_2O)_6]$ $[Ni(glyphosate)_2]$ complex. They found that the loss of all water molecules occurs below 300°C, corresponding to a mass loss of 28.2%, very close to the calculated value of 27%, and attributed that value to six water molecules.

Menelaou et al. (2009) also observed two other losses above 300°C, and attributed the first one to loss of the carboxylate moiety and half of the phosphonate group. The second loss was attributed to the decomposition of the organic chain. The total mass loss corresponded to 45.7%, stopping at 935°C. We calculated the total mass loss of the complexes we synthesized. For the 2GlyP:1Ni complex, as shown in Table 5, the total loss corresponded to 48.9%, stopping at 930°C. This is in close agreement with the results for the crystalline complex obtained by Menelaou et al.

Table 5. Mass loss and number of water molecules involved

Complex	Mass loss (wt. %) below 300°C	Number of water molecules	Total mass loss (wt. %)
2GlyP : 1Ni (II) pH = 8	28.87	6	48.90
1GlyP : 1Ni (II) pH = 8	10.60	2	32.80
2GlyP : 1Ni (II) pH = 4	21.14	4.4	52.24
1GlyP : 1Ni (II) pH = 4	23.30	4.8	54.10

X-Ray Absorption Spectroscopy Using Synchrotron Radiation

The XAS spectra of the solid Ni-glyphosate complexes were measured at room temperature, and presented similar patterns. Figure 4 displays the normalized XAS spectra, where a weak pre-edge peak at 8332 eV, assigned to the forbidden 1s → 3d transitions, is observed in the four spectra. The edge energy was 8341 eV for all spectra, and it was followed by intense white lines, assigned to allowed 1s → 4p transitions and strong oscillations in the post-white line region of the spectra, characteristic of Ni (II) in an octahedral environment (Colpas, 1991).

The k^3-weighted EXAFS spectra of the complexes in the 3 – 12 Å^{-1} range are shown in Figure 5. The spectra a and b (1:1 pH = 4 and 1:2 pH = 4, respectively) display a sinusoidal pattern, but it is possible to identify the emergence of a shoulder near 5 Å^{-1} in spectrum c (1:1 pH = 8, indicated by an arrow), which finally splits into two components in spectrum d (2:1 pH = 8).

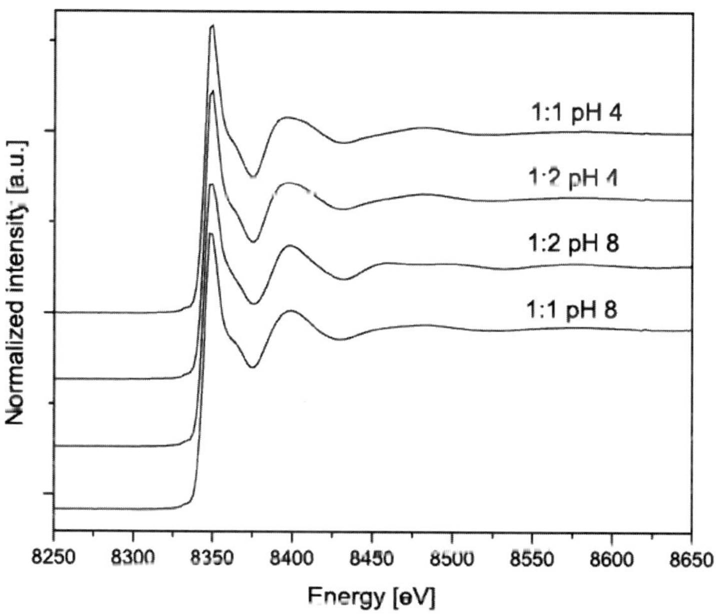

Figure 4. XAS spectra of the Ni(II)-glyphosate complexes measured in the solid state.

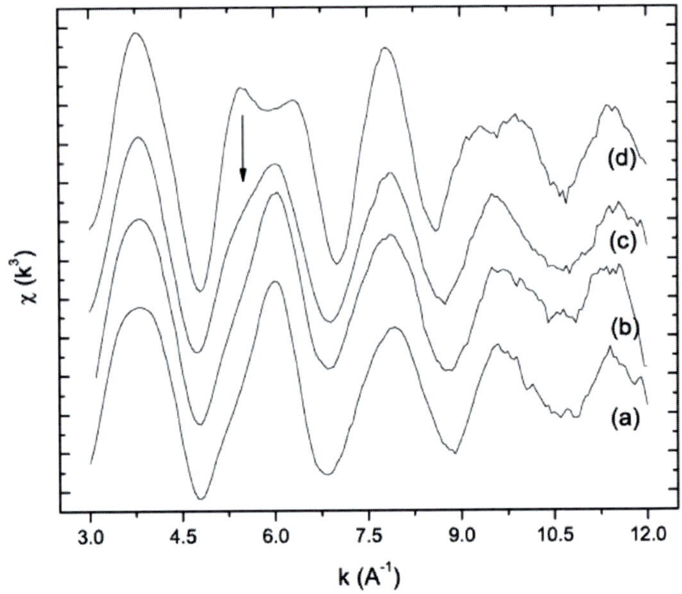

Figure 5. EXAFS spectra (k^3-weighted) of the Ni (II)-glyphosate complexes (a) 1:1 pH 4 (b) 1:2 pH 4 (c) 1:1 pH 8 (d) 1:2 pH 8.

Figure 6 shows the pseudo-distribution profiles of the radial functions (pdrf), obtained from the EXAFS spectra. These pseudo-distributions, or EXAFS profiles, represent the distances of the scattering atoms from the Ni (II) absorbing center.

The differences of the local environments around the absorbing Ni(II) atom is quite clear in the EXAFS profiles. We analyzed these differences as a function of the initial synthesis conditions. Each spectrum presents a characteristic strong peak between 1 – 2 Å. However, whereas the EXAFS profile of the complex obtained at 1:1 stoichiometric ratio and at pH = 4 presents this single peak, the profile of the complex obtained at the same pH value, but with excess glyphosate, presents in addition two other peaks above 2 Å and a broad peak near 3.5 Å.

For the profile of the complex obtained at 1:1 stoichiometric ratio and pH = 8, the peaks above 2 Å are more defined, with the exception of the region 3.5 Å.

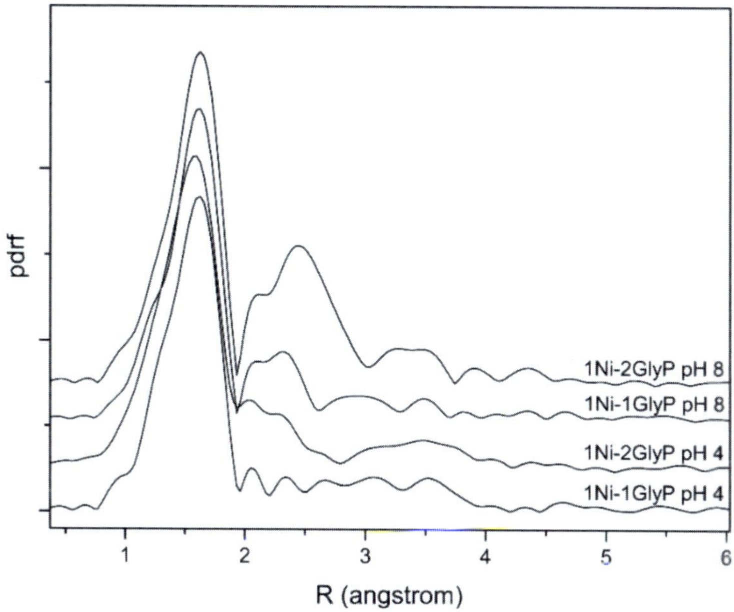

Figure 6. EXAFS profiles of the Ni-glyphosate complexes.

Finally, for the complex obtained at pH = 8, but with excess glyphosate, there are intense peaks above 2 Å, as well as in the 3.5 Å region. Based on the XANES spectra, as well as the electronic spectroscopy in the visible region and FTIR spectroscopy, we can infer that the first (and common to all EXAFS profiles) peak was due to the scattering of six light atoms of the first Ni(II) coordination sphere, presumably oxygen, or less probably, nitrogen atoms.

It should be noted that the EXAFS technique is not sensitive to light atoms, so it is not possible to determine if the atoms in the first coordination sphere are oxygen or nitrogen, nor the number of oxygen and nitrogen atoms. Such limitation of the technique is clearly revealed by the simulation of the first coordination sphere, using the FeFF6 code, as shown in Figure 7. In the simulation, we used six oxygen atoms, in octahedral geometry at a mean Ni-O distance of 1.96(1) Å, taken from crystallographic data (Menelaou, 2009).

In other words, based on the FeFF6 simulation results, the first coordination sphere of all complexes is composed of six light atoms. Since the EXAFS technique is insensitive to light atoms, it is not possible to discriminate between oxygen and nitrogen bonding to the Ni(II) absorbing center. However, the EXAFS spectra corroborate the FTIR results in the low frequency region.

Besides that, we should emphasize that the crystallographic data that were used as starting point for the simulation of the EXAFS spectra included two Ni(II) species, the anionic complex $[Ni(glyphosate)_2]^{2-}$ and the cationic complex $[Ni(H_2O)_6]^{2+}$, which might lead to dispersion in the Ni – O bonding distances.

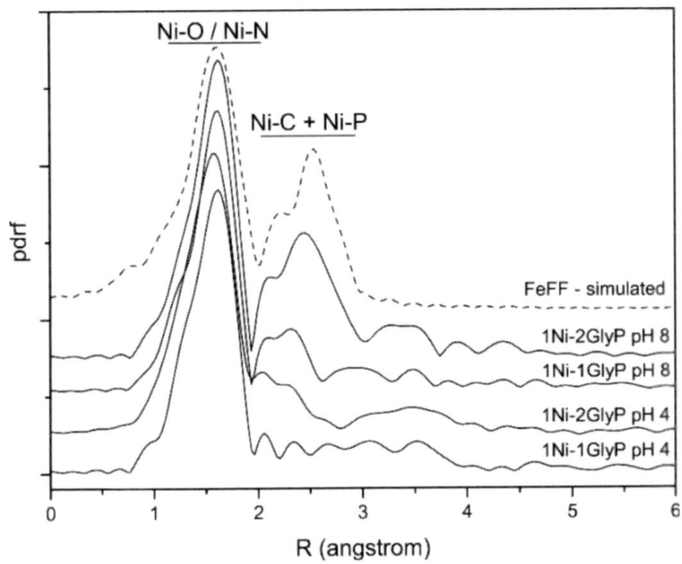

Figure 7. Comparison of experimental and simulated EXAFS profiles, obtained from the FeFF6 code.

Nevertheless, it is important to note that the EXAFS spectrum of the 1Ni : 1GlyP complex obtained at pH 4 was consistently simulated by using only one coordination sphere. On the other hand, the complex 1Ni : 2GlyP obtained at pH = 8 presented the richest EXAFS profile of the four complexes. Besides that, since previous characterization results suggest that that complex indeed is composed of $[Ni(glyphosate)_2][Ni(H_2O)_6]$, we carried out a deeper analysis of the EXAFS spectrum of this complex.

Table 6. Bond distances used for the simulations

Atomic pair * group	d [Å]
Ni(1)-O * COO-	2.05[a]
Ni(1)-N * NH	2.095
Ni(1)-O * POOHO-	2.12[a]
Ni(1)-P(1)	3.070
Ni(1)-P(2)	3.057
Ni(1)-C(1) *CH2	2.917
Ni(1)-C(2) * CH2	2.931
Ni(1)-C(3) * COO-	2.866
Ni(1)-C(4) * CH2	2.928
Ni(1)-C(5) * CH2	2.923
Ni(1)-C(6) * COO-	2.847
Ni(2)-O * H_2O	2.05[a]

[a] values were taken as mean distances.

The analysis involved simulations of the peaks observed above 2 Å in the EXAFS profile, by adding a second coordination sphere composed of six carbon atoms at a mean distance of 2.85(3) Å from the absorbing Ni(II) center, and a third coordination sphere, composed of two phosphorus atoms at a mean distance of 3.03(4) Å. Those distances, listed in Table 6, were taken from the CIF file of the crystal structure published by Menelaou et al. (2009), deposited in the Cambridge Crystallographic Data Centre (CCDC, file 722263).

As shown in Figure 7, the simulations of the EXAFS spectra of the complex 1Ni: 2GlyP obtained at pH = 8 using three coordination spheres had good agreement with the experimental spectrum. Besides that, the EXAFS profiles of the complexes 1Ni: 2GlyP (pH 4) and 1Ni : 1GlyP (pH 8) could not be satisfactorily simulated using higher coordination spheres, suggesting a high dispersion of the Ni(II) coordination sites. However, it should be remembered that the first coordination sphere was satisfactorily adjusted for all complexes, using an octahedral environment around the Ni(II) absorbing atom. Table 7 lists the adjusted parameters obtained for the two complexes.

The goodness-of-fit can be estimated by the minimum square adjustment between the experimental spectrum of the complex 1Ni: 1GlyP (pH 4) and its simulation in the region of the peak of the first coordination sphere, which affords R = 0.98. For the experimental spectrum of the complex 1Ni: 2GlyP (pH 8), the goodness-of-fit was R = 0.96.

In the literature, there are few examples of EXAFS studies on coordination complexes containing glyphosate. To our knowledge, only the systems Cu(II)-glyphosate and Cd(II)-glyphosate have been investigated by EXAFS (Sheals, 2001; Ramstedt, 2004).

Sheals et al. (2001) reported that the glyphosate molecule is three-coordinated to the Cu(II) center, bonded to the corners of a Jahn-Teller distorted octahedron, with an equatorial geometry. Completing the coordination, three water molecules are bonded to the Cu(II) center.

Table 7. Adjusted parameters obtained from the simulation of the EXAFS profiles

Complex	**Coordination spheres – single scattering**								
	1st (Ni-O/Ni-N)			2nd (Ni-C)			3rd (Ni-P)		
	N	R (Å)	σ^2	N	R (Å)	σ^2	N	R (Å)	σ^2
1:1 pH 4	6	1.96(1)	0.005	-	-	-	-	-	-
2:1 pH 8	6	1.96(1)	0.006	6	2.85(3)	0.007	2	3.03(4)	0.007

N = coordination number (fixed); R = Ni-scattering atom distance; σ^2 = Debye-Waller factor.

Ramstedt et al. (2004) investigated a Cd(II) glyphosate complex in solution. They determined the formation constants of a series of Cd (II)-glyphosate complexes by potentiometric titration, and the predominant species were also analyzed by EXAFS. They report different structures for the complexes obtained in 1:1 and 1:2 stoichiometric ratios.

Conclusion

Monitoring of the reactions by electronic spectroscopy in aqueous solution revealed that in the experimental conditions of this work, complexation of the glyphosate molecule to the Ni(II) cations occurs. In the solid state, the complexes were analyzed by FTIR in the 4000-150 cm^{-1} region, showing that in most cases the carboxylate groups of the glyphosate molecule are coordinated in a monodentate fashion. Coordination of glyphosate occurs through the phosphonate and amino groups, with water molecules participating in the coordination sphere.

Joining the FTIR results with those from elemental analysis allows us to conclude that in the 1:1 complexes obtained at pH = 8 for the glyphosate molecule are tetra-coordinated. The thermal analysis also indicated that water molecules participate in the coordination.

Simulation of the EXAFS spectra also agree with FTIR results, in the sense that the complexes obtained in acidic pH are formed by only one coordination sphere, composed of six light atoms, presumably oxygen and/or nitrogen, regardless the initial stoichiometry. On the other hand, the complex obtained in basic pH with excess glyphosate is composed by two complex units, an anionic $[Ni(glyphosate)_2]^{2-}$ and a cationic one, $[Ni(H_2O)_6]^{2+}$.

The results obtained for the complexes at pH = 8 can help understand the physiological behavior of glyphosate in the phloem medium.

References

Amarante Jr., O. P.; Santos, T. C. R.; Brito, N. M.; Ribeiro, M. L. *Quím Nova.* 2002, *25*, 589-593.

Barja, B. C.; Herszage, J.; Afonso, M. S. *Polyhedron*. 2001, *20*, 1821-1830.

Boyd, A. S.; Sommers, L. E.; Nelson, D. W. *Soil Sci Soc Am J.* 1981, *45*, 1241-1242.

Clarke, E. T.; Rudolf, P. R.; Martell, A. E.; Clearfield, A. *Inorg Chim Acta.* 1989, *164*, 59-63.

Colpas, G. J.; Maroney, M. J.; Bagyinka, C.; Kumar, M.; Willis, W. S.; Suib, S. L.; Baidya, N.; Mascharak, P. K. *Inorg Chem.* 1991, *30*, 920-928.

Cotton, F. A.; Wilkinson, G. *Advanced Inorganic Chemistry*; 4th Edn; John Wiley & Sons: New York, NY, 1980; pp 787.

Dubbin, E. W.; Sposito, G.; Zavarin, M. *Soil Sci.* 2000, *9*, 699-707.

Duke, S. O.; Lydon, J.; Koskinen, W. C.; Moorman, T. B.; Chaney, R. L.; Hammerschmidt, R. *J Agric Food Chem.* 2012, *60*, 10375-10397.

Han, F. G.; Luo, H. Z.; Ye, Q.; Xiong, R. G. *Z Anorg Allg Chem.* 2008, *634*, 1991-1995.

Heineke, D.; Franklin, S. J.; Raymond, K. N. *Inorg Chem.* 1994, *33*, 2413-2421.

Jing, C.; Yan, W. *Environ Sci Technol.* 2018, *52*, 1946-1953.

Kobylecka, J.; Ptaszynski, B.; Zwolinska, A. *Monatsh Chem.* 2000, *131*, 1-11.

Kutman, B. Y.; Kutman, U. B.; Cakmak, I. *J Agric Food Chem.* 2013, *61*, 8364-8372.

Liu, B.; Dong, L.; Yu, Q.; Li, X.; Wu, F.; Tan, Z.; Luo, S. *J Phys Chem B.* 2016, *120*, 2132-2137.

Malavolta, E.; Moraes, M. F. *Better Crops*. 2007, *91*, 26-27.

Menelaou, M.; Dakanali, M.; Raptopoulou, C. P.; Drouza, C.; Lalioti, N.; Salifoglou, A. *Polyhedron*. 2009, *28*, 3331-3339.

Nakamoto, K. *Infrared and Raman Spectra of Inorganic and Coordination Compounds, Part B.* 5th Ed. John Wiley & Sons: New York, NY, 1997; pp 59-62.

Peixoto, M. M.; Bauerfeldt, G. F.; Herbst, M. H.; Pereira, M. S.; da Silva, C. O. *J Phys Chem A.* 2015, *119*, 5241-5249.

Prenesti, E.; Gulmini, M. *Ann Chim.* 1998, *88*, 591-595.

Ramstedt, M.; Norgren, C.; Sheals, J.; Sjöberg, S.; Persson, P. *Inorg Chim Acta*. 2004, *357*, 1185-1192.

Sheals, J.; Persson, P.; Hedman, B. *Inorg Chem.* 2001, *40*, 4302-4309.

Subramaniam, V.; Hoggard, P. E. *J Agric Food Chem.* 1988, *36*, 1326-1329.

Theriot, L. J.; Carlisle, G. O.; Hu, H. J. *J Inorg Nucl Chem.* 1969, *31*, 2891-2894.

Undabeytia, T.; Morillo, E.; Maqueda, C. *J Agric Food Chem.* 2002, *50*, 1918-1921.

Zobiole, L. H. S.; Oliveira Jr., R. S.; Kremer R. J.; Constantin, J.; Yamada, T.; Castro, C.; Oliveira, F. A.; Oliveira Jr., A. *Appl Soil Ecol.* 2010, *44*, 176-180.

In: Glyphosate
Editor: Ron Quinn

ISBN: 978-1-53614-002-6

Chapter 4

Mechanisms of Toxicity of a Glyphosate-Based Herbicide in Mammals

Vilmair Zancanaro, MD, Emyr Hiago Bellaver, MD and Claudriana Locatelli*, PhD

University of Oeste de Santa Catarina – UNOESC, Videira, Santa Catarina, Brazil
University of Alto Vale do Rio do Peixe – UNIARP, Caçador, SC, Brazil

Abstract

Glyphosate-based herbicides (GBH) are the major pesticides used worldwide. Their uses have increased exponentially since their introduction on the market. Residue levels in food, water, and human urine has been detected, raising concerns about potential environmental and human health impacts. Research has shown that glyphosate can

* Corresponding Author Email: claudriana.locatelli@unoesc.edu.br; claudrilocatelli@gmail.com.

induce a broad range of biological effects in exposed organisms, particularly via generating oxidative stress. However, no comprehensive investigation has been undertaken of the global molecular mechanisms of toxicity of glyphosate for mammalian species. Studies show that glyphosate has an oral absorption of 20 to 40% and is considered to be of low toxicity to mammals; however, recent research has shown that it can induce serious damage to mammalian cells. Suggested mechanisms of toxicity include blocking of mitochondrial oxidative phosphorylation, inhibition of cytochrome p450 activity, inhibition of intestinal arylhydrocarbohydroxylase activity, changes in glucose 6-phosphate dehydrogenase activity, DNA damage, hormonal changes, channel openings of calcium, and neurotoxicity involving NMDA receptor activation. These cellular alterations are directly related to the long-term development of teratogenic, tumorigenic, and hepatorenal effects. Therefore, this chapter aims to demonstrate the mechanisms of toxicity involving glyphosate in mammals.

Keywords: glyphosate herbicide, neurotoxicity, hepatotoxicity, renal toxicity, hormonal changes

INTRODUCTION

An increased number of chemicals in natural ecosystems is one of the consequences of industrializing the agricultural sector. Large-scale use of agrochemicals has become a subject of study at physiological levels, since the side effects of pesticides used by the agricultural industry constitute an important risk factor for environmental health. This fact is responsible for awakening the need for a sustainable, ecological approach, one that has especially a lower impact factor on health and the environment, generating a new agricultural concept in relation to food production.

Exposure to agrochemicals occurs through occupational, environmental, and dietary routes, through skin contact, ingestion, and even inhalation of these compounds (Androutsopoulos et al., 2013; Cechhi et al., 2012; Pournourmohammadi et al., 2007). In 2015, there was a forecast of 220,000 deaths due to direct or indirect contact with pesticides. Contamination with pesticides is considered a public health problem according to the World Health Organization (WHO), which estimates that around three million people are infected each year worldwide (Nicolopoulou-Stamati et al., 2016; Ubessi et al., 2015; Cechhi et al., 2012). Some effects associated with the use and contamination of pesticides by humans include dermatological, gastrointestinal, neurological, carcinogenic, and respiratory tract pathologies, as well as reproductive and endocrine disorders (Androutsopoulos et al., 2013; NSW EPA, 2013; Hoffman; Capel; Larson, 2000; Baylis, 2000).

Among chemical pesticides, organophosphates (OPs) have appeared on the agricultural market as a greener alternative to organochlorines; among the range of pesticides in this class glyphosate is emphasized. Glyphosate is widely used, related to current agriculture, and nontoxic in form of salt, but is considered toxicogenic when associated with surfactants in commercial formulations such as Roundup (Mesnage et al., 2015; Kim et al., 2013; Baylis, 2000). Glyphosate [N-(phosphonomethyl) glycine] is a post-emergence herbicide that is non-selectively applied to control the growth of long grasses and broad-leafed weeds. Glyphosate is among the top pesticides used in the world (Mallat and Barceló, 1998). The widespread use of this compound is due to its low mammalian toxicity, because it destroys plants and microorganisms by inhibiting shikimate pathway; thus, it was considered to be completely nontoxic to animals and humans (Mesnage et al., 2015).

Glyphosate inhibits the activity of the enzyme synthase 5-enolpyruvylshikimate-3-phosphate synthase (EPSPS) by competitive inhibition, which prevents the production of chorismate, a precursor for aromatic amino acids used in the synthesis of a number of pigments, flavonoids, and anthocyanins (Gomes et al., 2014; Ribeiro et al., 2014). Thus, glyphosate was not thought to have a common molecular target in animal species (Sandrini et al., 2013). However, mounting evidence suggests that non-target species may also be affected (Giesy et al., 2000).

Approximately 30% of glyphosate is orally absorbed; in contrast, its dermal absorption is poor (Bradberry et al., 2004). The herbicide is primarily excreted unchanged in urine, although aminomethylphosphonic acid (AMPA) (Figure 1) is a recognized minor metabolite in humans (Hori et al., 2003).

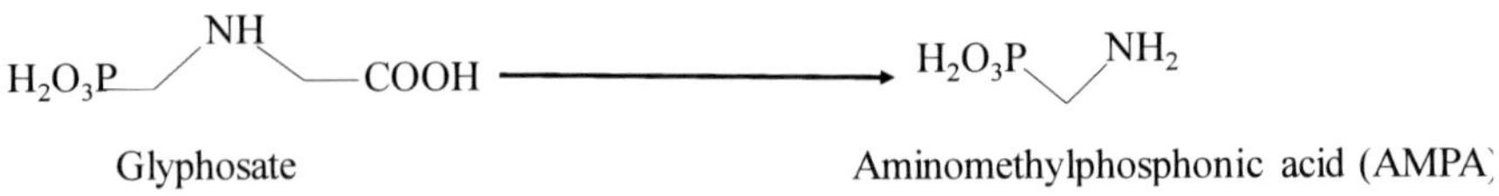

Figure 1. Chemical structure of glyphosate, its metabolites – aminomethylphosphonic acid (AMPA).

Glyphosate was already detected in the urine of children, mothers, and fathers, at concentration levels up to 5.0 ng/ml in the urine of non-farm individuals and up to 11 ng/ml in the urine of farm individuals; these results may be linked to the environment and to exposure due to diet (Curwin et al., 2007).

Contamination by glyphosate is related to: hepatic enzyme instability; an interruption of glycemic homeostasis and lipid metabolism (Cecchi et al., 2012; Karami-Mohajeri and Abdollahi, 2011; Pournourmohammadi et al., 2007); alteration of genetic transcription factors and DNA fragmentation (Androutsopoulos et al., 2013; Everett and Matheson, 2010; Shadnia et al., 2005); changes in maternal-placental-fetal balance (Cecchi et al., 2012); and genotoxic effects on mitochondrial function, triggering oxidative stress in the cells and consequently pathologies of the nervous and endocrine systems (Myers et al., 2016; Mesnage et al., 2015; Karami-Mohajeri and Abdollahi, 2011).

Glyphosate contains phosphorus but does not inhibit acetyl-cholinesterase in humans (Bradberry at al., 2004). In animal studies, it has been suggested that glyphosate may uncouple mitochondrial oxidative phosphorylation, although it is more likely for the surfactant to interfere with mitochondrial function by damaging mitochondrial membranes (Bradberry et al., 2004). The mechanism of toxicity of glyphosate formulations in humans is unclear. Ingestion of small amounts of glyphosate may have mucocutaneous side-effects, such as oral ulceration, oesophageal symptomatology, hypersalivation, nausea, vomiting, and diarrhea (Chen et al., 2009; Lee et al., 2008). Ingestion of larger amounts (usually >85 mL) causes significant toxicity, including renal and hepatic impairment, acid-base disturbance, hypotension, and pulmonary edema (Chen et al., 2009; Lee et al., 2008).

McQueen et al. (2012) showed in their study that the exposure of pregnant women to glyphosate is less than 2% of the acceptable daily intake, and they estimate that only 15% of this amount crosses the placenta (McQueen et al., 2012). The placenta is very important in intrauterine development; its role involves controlling growth and substance exchange between the fetus and the mother (Aris and Leblanc, 2011). Richard et al. (2005) have shown that a nontoxic (below the recommended) amount of glyphosate was still toxic for the human placental cell line, and that it also disturbed estrogen production. Benachour et al. (2007) showed in their study that glyphosate has a toxic and endocrine-disrupting effect in human embryonic cells (in lower nontoxic concentration) and in placental-derived cells. Fresh human placental cells were also sensitive to lower doses of glyphosate, showing endocrine disruption characteristic of this substance, consequently affecting human reproduction and fetal development (Benachour et al., 2007). Furthermore, in 2015, the WHO International Agency for Research on Cancer reclassified glyphosate-based herbicides (GBH) as “probably carcinogenic to humans” (Fritschi et al., 2015). In addition, studies on the gonadal axis have shown that glyphosate interferes with the activity of aromatase, leading to changes in reproductive development in rats (Romano et al., 2012, 2010) and thus demonstrating the potential for endocrine disruption through GBH.

Impaired consciousness and seizures have also been reported as sequelae, but information is limited on the effects of glyphosate toxicity on the central nervous system (CNS). Post-mortem analysis of one patient detected glyphosate in brain tissue, which indicated transfer across the blood-brain barrier (Chang et al., 2009; Menkes et al., 1991). In this context, it was reported that acute and chronic exposure to glyphosate might cause Parkinsonism, a condition similar to Parkinson’s disease (Wang et al., 2011; Barbosa et al., 2001).

In this context, epidemiological studies in experimental animals show that the toxicity of glyphosate is related to hepatic, renal, hematological, and hormonal alterations, as well as damages to the central nervous system. These studies also report that these damages appear to be related to the ability of glyphosate to induce oxidative stress. Therefore, this chapter aims to show the mechanisms of hepatic, renal, hematological, hormonal, and neural toxicity associated with mammalian exposure to glyphosate.

The Action of Glyphosate on the Liver and Renal System

The liver and the kidney are the two major organs that suffer from oxidative stress, since glyphosate is mainly metabolized in the liver and the metabolites are discharged into the kidney (Li et al., 2017).

The hepatic effects of glyphosate were first observed in 1980, in its ability to alter mitochondrial oxidative phosphorylation of the liver (Olorunsogo; Bababunmi; Bassir, 1979). Glyphosate can act by increasing the permeability of the mitochondrial membrane to the protons, and Ca^{2+} (Olorunsogo, 1990) can trigger the production of reactive oxygen species (EROs), resulting in oxidative stress (Cavalli et al., 2013). Elevation in oxidative stress markers was detected in the livers and kidneys of rats after subchronic exposure to glyphosate in the United States, using a glyphosate concentration of 700 μg/L in drinking water (Larsen et al., 2012). Hepatic histological biochemistry changes were detected in rats consuming 4.87 mg of glyphosate per kg of body weight every 2 days for 75 days (Benedetti et al., 2004).

To confirm the damage caused by glyphosate in the liver of rats, Li et al. (2017) determined serum levels of the major liver enzymes, Aspartate Amino Transaminase (AST) and Alanine Amino Transaminase (ALT). The results showed that serum AST and ALT levels significantly increased with a treatment of 500 mg/kg glyphosate, compared to the control group ($p<0.05$). These results showed that glyphosate can affect hepatic metabolism, causing oxidative damage to liver tissue. The authors observed a significant decrease in total superoxide dismutase (SOD) activity and increased malondialdehyde (MDA) content in the serum, liver, and kidneys in the 500 mg/kg group, indicated the presence of oxidative stress. In addition, increased levels of hydrogen peroxide (H_2O_2) and catalase (CAT) activity in serum and liver, and decreased glutathione (GSH) and glutathione peroxidase (GSH-Px) activity in renal tissue, also confirmed the occurrence of oxidative stress.

Mesnage et al. (2015) conducted a two-year study in rats, administering 0.1 ppb Roundup (50 ng/L glyphosate equivalent) through drinking water (with a daily intake of 4 mg/kg body weight/day of glyphosate). A higher incidence of anthropomorphological and biochemical changes in blood and urine were indicative of hepatic and renal functional pathology. Genetic disorders associated with chronic low-dose administration reflected a lipotoxic liver and kidney condition, along with increased cell growth that may be associated with regeneration in response to toxic effects that cause tissue damage. The results suggest that chronic exposure to glyphosate in a model laboratory toxicity system at an ultra-low environmental dose may result in damage to the liver and kidneys, with potentially significant implications for the health of animal and human populations.

The toxicity of GBH in goldfish (*Carassius auratus*) was evaluated by Li et al. (2016) and Li et al. (2017) after long-term exposure. Brain tissues, kidneys, and livers were collected and analyzed. Prolonged exposure to glyphosate caused disorders of blood biochemical indices and damage to renal tissue. Significantly-reduced levels of SOD and GSH-Px were seen, as well as elevated levels of MDA in serum. As the main antioxidant enzyme system in the mitochondrial matrix, SOD and GSH-Px can react directly with free radicals to eliminate damage caused by reactive oxygen species. Reduced levels of these antioxidants have demonstrated that exposure to glyphosate destroyed the antioxidant defense system of goldfish (Han et al., 2015) (Figure 2).

In order to explore the effect of occupational exposure to glyphosate on hepatic and renal function, Zhang et al. (2017) evaluated 526 workers who were occupationally exposed to glyphosate from five glyphosate-producing factories, along with 442 administrative staff who were not exposed to glyphosate and were selected as controls. The research concluded that glyphosate may affect liver and kidney function among the exposed population, and there was an association between the effect and the dose exposure. The difference between the exposed population group and the control group was statistically significant, and abnormal renal and hepatic function rates were 8.0% and 9.8%, respectively.

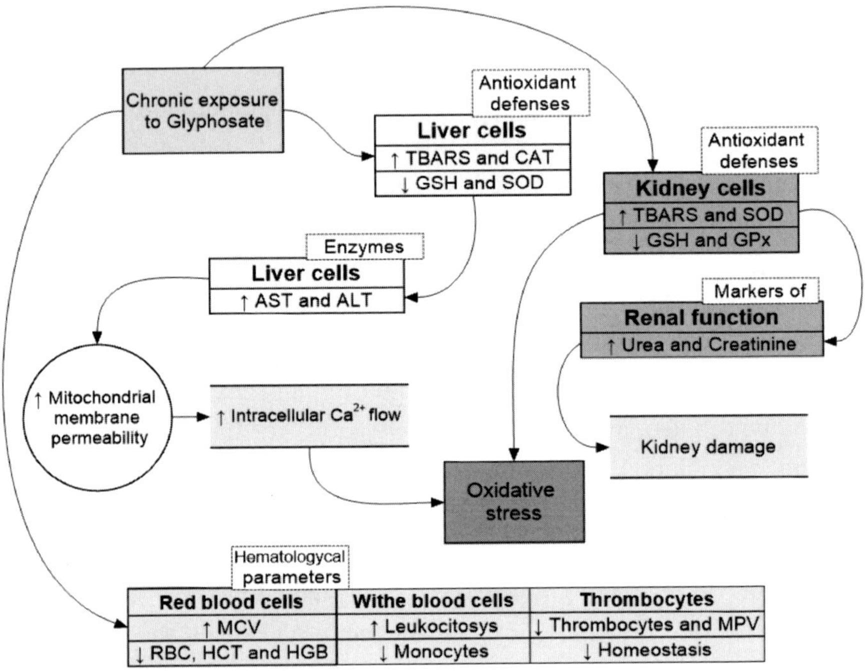

Figure 2. The action of glyphosate after chronic exposure to hepatic, renal, and hematological tissue cells.

Indirakshi et al. (2017) reported on a patient who consumed 15 mL of glyphosate, resulting in toxic epidermal necrolysis and acute renal injury. The results of the tests showed: serum creatinine of 10.1 mg/dL, urea of 201 mg/dL, hemoglobin of 13.2 g/dL, total leukocyte count of 6,800/mm^3, platelet count of 60,000/mm3, total bilirubin of 0.7 mg/dL, serum AST of 77 U/L, and ALT of 46 U/L. The patient underwent eight hemodialysis sections, and after 3 months, serum creatinine and blood urea were 1.0 and 24 mg/dL, respectively. The greater amount of glyphosate is excreted unchanged in the urine, and the data in humans is limited (Bradberry; Proudfoot; Vale, 2004). The patient had no gastrointestinal cause or hypotension due to any other reason, and thus no other way to explain the acute renal injury. The patient recovered completely from acute renal injury, highlighting a possible role of hemodialysis in the management of glyphosate intoxication. Serum levels of glyphosate could not be achieved.

THE ACTION OF GLYPHOSATE ON HEMATOLOGICAL PARAMETERS

Hematological studies aim to investigate the patterns and alterations of blood cells, since blood characteristics are related to the adaptability of animals to the environment. Therefore, the analysis of hematological characteristics is of great importance in animals exposed to chemicals (Lopes et al., 2010).

Hematological parameters may change in response to xenobiotics, as mentioned by Modesto and Martinez (2010). Seralini et al. (2014) reported the results of a long-term study in rats throughout their lifespan. Water-treated animals containing Roundup® (0.1 parts per billion) or Roundup®-treated transgenic maize presented about 70 statistical differences in hematological, urinary, organ weight, body weight, and feed consumption parameters. These changes increased the risk of developing damage to the gastrointestinal system, kidneys, and liver.

Jasper et al. (2012) showed that the mean values of the parameters decreased in Roundup®-treated animals at a dose of 500 mg/kg body weight for 15 days; specifically, there was a significant reduction in erythrocyte (RBC) numbers, hemoglobin (HGB) concentration, and hematocrit (HCT), with and increase in mean corpuscular volume of the red blood cell (MCV), characterizing macrocytic anemia. Pieniazek et al. (2004) observed significant lysis in the blood of glyphosate-treated Wistar animals compared to controls (untreated). According to Benedetti et al. (2013), exposure to glyphosate demonstrated reflexes in peripheral blood hematological data with a reduction in erythrocyte numbers, hemoglobin concentration, and hematocrit (Figure 2).

In a study conducted by Chan and Mahler (1992), rats did not show significant changes in platelet counts after administration of pure glyphosate in feed at a concentration of 50,000 ppm/kg of feed, although the animals presented signs of dehydration which may have compromised the results.

When considering the bleeding events observed in rats, Neiva et al. (2010) investigated the possible in vitro effects of glyphosate on platelets and blood coagulation. In conclusion, the authors demonstrate that glyphosate caused changes in platelet metabolism, with an inhibitory effect on hemostasis.

Hematological parameters (leukocytes, monocytes, and platelets) were identified by Aroonvilairat et al. (2015), who evaluated farmers involved in orchid cultivation in Thailand. However, in their research, reductions in hematological variables were not statistically significant.

García-García et al. (2016) and Wafa et al. (2013) justified divergences in comparisons of hematological parameters between groups by alleging disruptive action of pesticides in the hematopoietic system, inducing adaptive reactions of the bone marrow to oxidative damage caused by pesticides. In addition to a reduced number of platelets, the mean platelet volume (MPV) was lower in the group that used pesticides. A similar result was found by Varol, Ogut and Gultekin (2014), who evaluated platelet variation and its indices in workers exposed to pesticides. This study verified a significant reduction in the number of platelets and in the MPV, suggesting that MPV could be an indicator that is sensitive to the effect of exposure to pesticides.

Cattelan's (2017) study held the objective of finding possible associations between the use of pesticides on biochemical, hematological, and oxidative stress parameters in a group of rural workers from Santiago, Rio Grande do Sul, Brazil. Cattelan (2017) concluded that the profile of alterations in serum levels of alkaline phosphatase, albumin, total cholesterol, total leukocytes, monocytes, and platelets is related to the use of pesticides, suggesting a chronic effect caused by their use in the working environment under the conditions evaluated. Lower levels of antioxidant enzymes (SOD, GPX, and GSH) have demonstrated that organisms exhibit a physiological response to such external agents.

These divergences may also be associated with different types of agrochemicals used. The progression of hematotoxicity induced by pesticides and their mechanisms of action in hematopoiesis is not well explained (Chatterjee et al., 2013), and numerous animal studies have attempted to reproduce the exposure conditions and mixtures of pesticides in order to establish mechanisms and metabolic pathways of toxicity of pesticides and their mixtures with the human organism (Hadrup et al., 2016; Porreca et al., 2016; Chatterjee et al., 2013).

Glyphosate-Based Herbicides Cause Disruption in Organs and Endocrine Systems

Pournourmohammadi et al. (2007) and Anadon et al. (2009) reported that glyphosate phosphorylate affected the enzyme acetylcholinesterase (AChE) in nerve endings, resulting in enzymatic overstimulation by excess ACh (acetylcholine) in the same (Wang and Ho, 2012; Anadon et al., 2009; Pournourmohammadi et al., 2007; Glusczak et al., 2007), although there is no significant evidence of this inhibition in mammals. In the studies, the authors investigated the effects of the pesticide on insulin secretion in pancreatic islets by in vivo methodologies, and showed that the increase of glucose levels in the serum route contrasts the inhibition of acetylcholine (an important stimulator of insulin and glucagon production) in the islets of Langerhans, as a result of pesticide intoxication. Hyperglycemia was justified by the influence exerted by the glyphosate in the brain, skeletal muscle, and liver that end up stimulating the production of glucose (Pournourmohammadi et al., 2007; Rahimi and Abdollahi, 2007). The stimulation of the sympathetic nervous system during physiological stress triggered by long-time chronic exposure to pesticides leads to the release of catecholamines, glucagon, and growth hormone, resulting in the physiological events of glucose release to the serum and insulin resistance resulting in hyperglycemia (Rahimi and Abdollahi, 2007; Husain and Ansari, 1988).

The oxidative stress triggered by glyphosate results from the increase in reactive oxygen species and formation of nitrosamine, causing lesions in the cell membrane and interfering in mitochondrial activity, which may interfere in hypothalamic-pituitary-adrenal axis homeostasis and the autonomic sympathetic nervous system and lead to hyperglycemia events (Rahimi and Abdollahi, 2007; Mechanick, 2006). Oxidative stress caused by glyphosate in the liver leads to the interruption of glucose homeostasis in the level of muscle gluconeogenesis, glycolysis, and glycogenolysis, in addition to affecting the metabolism of lipids, proteins, and in the control of homeostasis of thyroid hormones. When activating the protein kinase C (PKC), which is essential in regulating the expression of cellular genes, the receptors and the channels of ions are present in the plasma membrane, among other activities (Cavalli et al., 2015). The hydrolysis of phospholipids leads to an increase in the free concentration of cytosolic calcium due to the action of ACh, resulting in the exocytosis of the insulin granules. Although insulin, glucagon, and Langerhans islet glucokinase (GK) levels are elevated, the sensitivity of islets decreases when stimulated by glucose (Ortiga- Carvalho et al., 2016; Karami-Mohajeri and Abdollahi, 2011; Panahi et al., 2006; Duttaroy et al., 2004). The induction of insulin resistance due to interrupting the mitochondrial function of the pancreatic islets leads to a state of hyperglycemia, which also seems to be related to the oxidative stress exerted by pesticides (Karami-Mohajeri and Abdollahi, 2011; Rahimi and Abdollahi, 2007; Pournourmohammadi et al., 2005) and to the decrease in availability of Glut-4 receptors in some cells (De Souza et al., 2017).

While mapping hepatic targets affected by glyphosate in rats using proteomics, Ford et al. (2017) reported that glyphosate is metabolized to glyoxylate by reacting with cysteine amino acids in liver proteins (Mesnage et al., 2017; Shadnia et al., 2005). Glyoxylate is a reactive glycosylating agent that causes lysis of the amine group of amino acids, peptides, and proteins, generating a kind of carbonyl stress in the presence of oxidizing agents such as superoxide anion and hydrogen peroxide.

Glyoxylate inhibits the oxidation of fatty acids, while glyphosate increases the percentage of hepatic cholesterol (Samsel and Seneff, 2015; Wangh and Ho, 2012; Anadon et al., 2009).

Hepatic mitochondrial oxidation generates an increase in the C92 enzyme responsible for the oxidation of mitochondrial fatty acids Acaa2 and C94, which are involved in the biosynthesis of ketone bodies. Treatment with glyphosate in high doses causes deficiencies in these thiolases and in the oxidation pathways of fatty acids, causing liver dysfunction and leading to an increase in triglyceride levels and storage of hepatic cholesterol. This results in episodes of hepatic steatosis, since the fatty acids that have not been oxidized may proceed to other pathways of lipid metabolism (Ford et al., 2017).

The action of glyoxylate as a product of the endogenous degradation of glyphosate seems to be related to the induction of cancer in cases of increased exposure to the pesticide, leading to believe that this mechanism is related to the decrease of bioavailability of vitamin E (Anadon et al., 2009).

Studies have shown a relationship between organophosphorus (OPs), such as glyphosate, and uterine and endocrine system disorders, such as metabolic disorders in the thyroid, the adrenal glands, and seminal vesicles (De Souza et al., 2017; GM Myers et al., 2016; Watch, 2016; Cecchi et al., 2012). However, the doses necessary for the induction of these disorders remain uncertain, since some laboratory findings considered abnormal at low doses cannot be seen in larger doses (Vandenberg et al., 2012; Casals-Casas and Desvergne, 2011), thus bringing to the fore the paradigm that "the dose does the poison". Some studies report that doses close to those acceptable for human exposure may be sufficient (Myers et al., 2016; Thongprakaisang et al., 2013). In contrast, other lines of research indicate cellular disorders due to oxidative stress mechanisms occurring in low doses (Thongprakaisang et al., 2013; A Polyzos et al., 2012; Schug et al., 2011). Still others show cellular disorders in overestimated doses that is, much higher than levels of human exposure (0.3mg/kg/day in the European Union and 1.75-2mg/kg body weight/day in the USA) (Varayoud et al., 2017; GM Watch, 2016; Mesnage et al., 2015).

The GBH disrupts the uterine development of female rats at 7-day exposure with doses of 2 mg/kg body weight/day, a reference dose considered safe for consumption of pure glyphosate in the USA. In females exposed to the pesticide after birth, structural changes were detected due to endometrial hyperplasia and interruption of proteins responsible for the organogenic differentiation of the organ (Schimpf et al., 2017). Varayoud et al. (2017) showed that GBHs are capable of modulating genes of uterine estrogen-sensitive receptors, such as ERα, Erβ, and PR, demonstrating that GBHs are capable of decreasing the active isoform of estrogen, ERα, in the epithelial cells of the uterus, whereas in the stroma the expression increased. The ERβ isoform, related to a possible defense mechanism against the undesirable effects induced by ERα, is present in almost all benign and malignant endometrial proliferative diseases, and was found to be elevated in uterine glandular cells. Since these ERs are central to normal uterine function, their dysregulation would be associated with uterine disorders like endometriosis or carcinomas (Schillace et al., 2014).

The increased expression of nuclear receptors for progesterone (PR) in the glandular compartment of rat uterus was induced by GBHs; such effects may affect normal development as well as uterine differentiation, affecting their fertility (Schimpf et al., 2017; Varayoud et al., 2017).

Perinatal exposure to GBHs leads to a change in the hypothalamic-pituitary-thyroid axis (HPT), causing an imbalance in thyroid hormone homeostasis in adult male rats. With of low dose exposures of GBH the experimental group showed decreased TSH levels but unaffected thyroid hormone (TH), T3 and T4 hormones in the serum, when compared to the control group, showing that the pituitary gland is more sensitive to a lower concentration of glyphosate. This unusual result in the hormonal profile may reflect changes in the TSH setpoint due to alteration of HPT genes during the fetal period, leading to hypothyroidism-compatible genetic frameworks in adult rats (De Souza et al., 2017).

Beecham and Seneff (2016) raised hypotheses about the use of GBHs and autism. The mechanism of action of glyphosate as a manganese ion chelator interferes in the activities of cells dependent on these ions decreasing to phosphatase 1 (PP1) proteins, TRH receptor-bound phosphate removers allowing the release, and TSH generating from the pituitary gland. This results in a decrease in serum TSH levels in pregnant women exposed to the pesticide during pregnancy, which is statistically related to autism in their offspring, since the fetus is totally dependent on maternal thyroid hormones for their neurological development (Jiang, Dias & He, 2014).

Exposure to GBHs during the gestational period also leads to disorders related to morphology and testicular function in the offspring. In addition, acute exposure to pesticides increases aromatase levels, produces molecular changes in the testicular barrier, and decreases normal sperm morphology, as well as affecting the functioning of Leydig cells in the testicles and affecting the production of estrogens such as testosterone (Varayoud et al., 2017; Romano et al., 2012; Walsh et al., 2000).

The role that Leydig cells play in male physiology is highly important. In a study using Leydig's MA-10 tumor cell model, glyphosate was able to disrupt the expression of the StAR protein, which is responsible for mediating the transfer of cholesterol from the outer membrane to the inner mitochondrial membrane, where the enzymatic molecule P450scc will cleave its side chain. This initiates the synthesis of steroid hormones, thus inhibiting steroidogenesis in these cells, affecting the fertility of the population exposed to the pesticide (Miller et al., 2015; Walsh et al., 2000).

Acute exposure to Roundup at doses of 0.036g/L for 30 minutes was able to induce oxidative stress in Sertoli cells in prepubertal testes of rats. GHB increased the calcium influx to the cellular interior; the intracellular calcium overload triggered the process of oxidative stress and cell death due to necrosis. Romano et al. (2010) and Clair et al. (2012) reported that the Roundup commercial formulation is capable of lowering testosterone levels and reducing re-testicular epithelium; its lumial diameter induces necrosis and apoptosis of testicular cells in adult rats. Cavalli et al. (2013) proposed that the toxicity of Roundup involves a calcium overload within the cell, a lack of cellular signaling, and low antioxidant defense; such mechanisms may have an impact on male fertility. On the other hand, in the tests conducted by Dai et al. (2016), the intake of 500 mg/kg body weight of glyphosate in pure salt was able to decrease the weight of the seminal gland and the total sperm count in the testes of rats. This fact was also found by Cai et al. (2017), although in the experiments carried out by Dai et al. (2016), pure salt was not able to trigger an oxidative stress process, much less influence the synthesis of hormones derived from cholesterol produced in the cells; the researchers concluded that glyphosate alone presents low toxicity in the reproductive system of animals.

By exposing human spermatozoa to 1 mg L of Roundup in order to analyze mitochondrial motility and integrity, Anifandis et al. (2017) concluded that such a quantity of pesticide exerted a deleterious effect on the motility of sperm cells. In addition, these cells presented a certain mitochondrial dysfunction, leading the authors to infer that this action was due to the glyphosate salt and not to the added surfactant in the commercial formulation. The probable mechanism of action involved in these results comes from the oxidative stress caused by the pesticide in the mitochondrial membrane, since this is essential in the production of cellular energy, hormonal biosynthesis, and integration of the functions of the steroid receptors (Amaral et al., 2013; Solakidi et al., 2005). Any alteration in its structure would lead to protein signaling apoptotic and consequent cell death.

Nardi et al. (2017) reported morphological disruptions in spermatozoa of prepubertal male rats at the level of spermatid decline and increased epididymal tail mass, decreased seminiferous tubules. Abnormal sperm morphology, the number of degenerate Sertoli and Leydig cells, and the hormonal level all increase with the decrease of testosterone, when the treatment of these animals with soy milk was supplemented with glyphosate in doses of 50 and 100 mg/kg. Changes in spermatozoa, as well as the decrease in testosterone levels, can be justified due to the destruction of Sertoli cells through a mechanism of cellular stress at the mitochondrial level, but this is a mechanism that is still not fully understood (Figure 3).

In view of the facts, it is possible to infer that the time, nature, and severity of impacts on the endocrine system may vary according to the degree of exposure to GBH, the means of exposure, age, and biological status, as well as different endogenous mechanisms that may affect the physiology of the organism. Studies corroborate that the different chronological levels of exposure to glyphosate-based pesticides will lead to serious complications over the course of days, months, and even years in *in vivo* and human experiments. Further research may elucidate the oxidizing mechanisms involved in the processes of disruption of endocrine homeostasis by GBH (Anifandis et al., 2017; Myers et al., 2016; Vandenberg et al., 2012; Diamanti-Kandarakis et al., 2009).

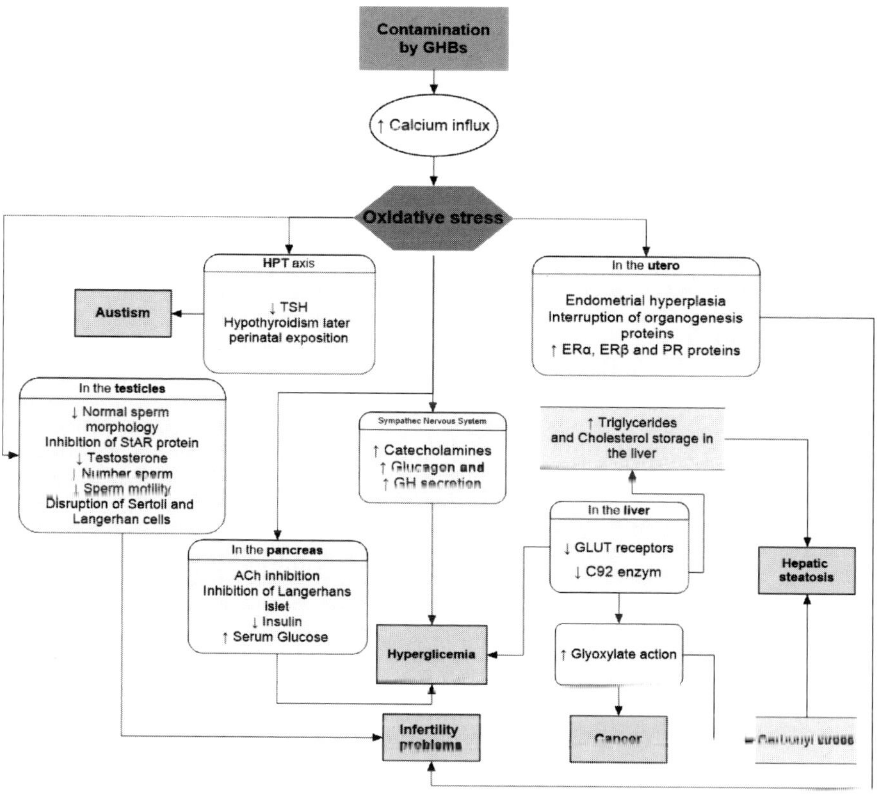

Figure 3. Summary of the effects of GHB on the organs, systems, and endocrine tissues latter oxidative stress.

Neurotoxicity Induced by the Exposition of Glyphosate-Based Herbicide

The low prevalence of neurological diseases, such as Parkinson's disease, in countries with low industrialization suggests that environmental factors associated with industrialization and the use of agrotoxic agents play a role in the etiopathogenesis of Parkinson's disease (Barbosa et al., 2001).

The clinical picture of severe glyphosate-based herbicides poisoning is manifested by gastroenteritis, respiratory disturbances, altered mental status, hypotension refractory to treatment, renal failure, and shock. A few case reports indicated possible neurotoxic sequels of glyphosate exposure with white matter lesions and development of Parkinsonism. Neurological involvement in particular extrapyramidal symptoms and signs including limb rigidity and resting tremor has only been reported following a few isolated events, rather than in the setting of chronic occupational exposure of glyphosate (Barbosa et al., 2001; Potrebic et al., 2009; Malhotra et al., 2010; Wang et al., 2011).

Barbosa et al. (2001) reported that a previously-healthy 54-year-old sustained chemical exposure in July 1995 while he was spraying glyphosate in a garden. He was wearing no protective gear such as gloves or a face mask, and the exposure to the chemical agent occurred as the breeze blew the spray back onto his trunk, arms, legs, and face. One month after the initial exposure, the patient displayed rigidity and slowness in all four limbs. One year later, he developed a resting slow tremor in the left hand and arm, and complained of impaired short-term memory. Family history was unremarkable. The physical examination performed under the influence of medication revealed a Parkinsonian syndrome. Potrebic et al. (2009) reported another case of intoxication by ingestion of 500 ml of the herbicide glyphosate that was associated with neurological alterations compatible with Parkinson's disease.

Malhotra et al. (2010) reported other cases of intoxication by glyphosate. One case included a 71-year-old male who was found unresponsive at home with an empty bottle of herbicide. This patient demonstrated prolonged clinical unresponsiveness (>7 days) following GBH poisoning. The mechanism of CNS toxicity of GBH is unknown, but possibilities for the prolonged period of unconsciousness (as seen in this patient) include encephalopathy due to the delayed effects of reduced organ perfusion and/or a hyperosmolar state and/or metabolic acidosis and direct neuronal toxic effects of the glyphosate–surfactant combination itself.

A previously-healthy 44-year-old woman presented with rigidity, slowness, and resting tremor in all four limbs, with no impairment of short-term memory, after sustaining long-term chemical exposure to glyphosate for 3 years as a worker in a chemical factory. Physical examination revealed a Parkinsonian syndrome. There was no known family history of neurological or other relevant disorders. The patient had consumed no other medications or herbal preparations before the onset of symptoms. Resting tremor in the left and right limbs, global akinesia, and rigidity were observed, but without a "cogwheel phenomenon" or postural instability (Wang et al., 2011). The mechanism by which glyphosate induced a Parkinsonian syndrome is unclear; however, studies in patients suggest that this is due to an excitotoxic mechanism (Barbosa et al., 2001; Brandberry et al., 2004; Malhotra et al., 2010).

Impaired consciousness and seizures have also been reported as sequelae, but data of glyphosate on CNS toxicity is limited. However, *in vitro* studies have revealed that glyphosate can pass through the blood-brain barrier and the placental barrier. Evidence is accumulating that organophosphate pesticides target serotonin and noradrenergic systems, contributing to adverse outcomes related to emotional and social behaviors (Aldridge et al., 2003; Meyer et al., 2005; Moreno et al., 2014; Negga et al., 2012).

Some work on rodents and cell culture indicate that exposure to glyphosate may contribute to important neurological changes. Coullery et al. (2016) used primary cultures of hippocampal pyramidal cells from Wistar rats at embryonic 18 days, who were treated for 1, 2, and 4 days with different concentrations of glyphosate (2, 4 and 6 mg/ml). This enzymatic test was based on the cleavage of 3-(4,5-dimethylth- iazol-2-yl)-2,5-diphenyl tetrazolium bromide (MTT, Sigma–Aldrich) into a blue-colored product (formazan) by the mitochondrial enzyme succinate dehydrogenase and was used to evaluate neurons cell viability. The results showed that the lower concentrations (2 and 4 mg/ml of glyphosate) did not affect neuronal survival at the evaluated periods. However, neurons exposed to the highest glyphosate concentration (6 mg/ml) showed a lethal effect after.

Coullery et al. (2016) also verified that glyphosate affects neuronal differentiation and development. Results revealed that glyphosate treatment induced a significant decrease in the number of polarized neurons. Thus, while around 70% of neurons polarized in control conditions, after glyphosate treatment only 38% of neurons differentiated, showing a discernible axon. These authors evaluated whether glyphosate induces a delay in axon differentiation after 2 days of treatment. The results revealed that even though glyphosate-treated neurons polarized, showing an axon labeled with Tau marker, the morphology of that axon was different from controls. Quantifications showed that around 80% of control neurons polarized after 48 h in culture, and at that time most glyphosate-treated neurons also polarized, since 70% of cells showed a clear axon (Tau+). These findings suggest that the herbicide treatment postpones neuronal differentiation, since treated neurons generated their axons later than controls. When neurons were exposed to glyphosate for 2 and 4 days, it was observed that control neurons developed a long and highly-branched axon, while after glyphosate exposure they elicited a simple axon, shorter and less branched than controls. These results suggest that glyphosate treatment induced a delay in axonal differentiation.

The delay in neuronal development induced by glyphosate could be related to a defect in neuronal connectivity and synaptic function in mature neurons. Consider the glyphosate effect on neuronal polarization; the morphology of the growth cones were evaluated in control and in treated undifferentiated neurons. The findings showed that treated neurons developed several and identical growth cones, which were significantly smaller than controls. The other observations suggest that changes in the actin organization in growth cones induced by glyphosate can postpone the ability of neurons to polarize and generate a clear axon (Coullery et al., 2016). The treatment of primary cultures of the hippocampal pyramidal cell by glyphosate led to a decrease in Wnt5a level a key factor for the initial neurite development and maturation as well as inducing a downregulation of CaMKII activity. This data suggests that morphological defects would likely be a consequence of the decrease in both Wnt5a expression and CaMKII activity induced by glyphosate axon (Coullery et al., 2016).

Studies were carried out by Ji et al. (2018) in the prefrontal cortex of mouse offspring exposed to glyphosate during pregnancy and lactation; they showed that miRNA expression is involved in glyphosate neurotoxicity. It is widely acknowledged that miRNAs can regulate target mRNAs and affect the activities of cells and tissues, implicating the importance of target gene prediction (Andrés-León et al., 2017). The results obtained by Ji et al. (2018) showed *Nr4a2* and *Wnt7b* were downregulated, while *Dkk1*, *Dixdc1, Runx1, Shh, Lef-1,* and *Axin2* were upregulated in the prefrontal cortex of mouse offspring. This finding indicated abnormalities of the Wnt/β-catenin and Notch pathways. These genes' expression changes may be associated with abnormal neural differentiation, and can contribute to synaptic degeneration during early stages of neurodegenerative disease.

In vivo studies conducted by Martínez at al. (2018) investigate the effects of glyphosate on CNS monoaminergic neurotransmitter contents in male Wistar rats treated with oral administration of glyphosate in doses of 35, 75, 150, and 800 mg/Kg/day for 6 days. The results showed that exposure to glyphosate, in a region and dose-dependent manner, was accompanied by a significant decrease in the serotonin, dopamine, and norepinephrine contents in the brain regions studied (striatum, hippocampus, prefrontal cortex, hypothalamus, hypothalamus, and midbrain). This indicated that glyphosate transfer across the blood-brain barrier enters the brain, probably accumulates in significant quantities, and exerts neurotoxicity, altering the serotonergic, dopaminergic, and noradrenergic systems.

Although studies have reported that glyphosate is an acetyl-cholinesterase-inhibiting organophosphate, Cattani et al. (2017) showed a decrease in acetylcholinesterase activity in the hippocampus of immature rats and adults exposed to GBH. However, they did not observe any change in peripheral blood acetylcholinesterase activity.

Cattani et al. (2014, 2017) propose that the neurotoxicity mechanism of glyphosate is related to glutamate excitotoxicity. The authors suggest that glyphosate stimulates the release of glutamate in the synaptic cleft. The increase of glutamate promotes a hyperactivation of the NMDA receptor with increased calcium influx. The intracellular increase in calcium concentration may be related to an increase in the production of reactive species, promoting an imbalance in the antioxidant defense system. The imbalance in the antioxidant system, in turn, may induce an overactivation of ERK 1/2 proteins associated with decreased phosphorylation of NFκB (Figure 4). Another relevant change found by Cattani et al. (2017) in immature rat and adult astrocytes after exposure to glyphosate was the decrease in serum levels of S100B in immature animals, and the increase in serum levels in adults. Changes in serum levels of S100B reflect changes the astroglial function. A recent pilot study demonstrated that S100B may serve as a marker for predicting neurologic complication in patients with glyphosate or glufosinate poisoning (Lee et al., 2017). These changes could be responsible for neural cell death.

In an attempt to summarize the findings described, a figure with effects of GBH in different tissues was elaborated (Figure 5), showing the different mechanisms of toxicity induced by glyphosate in mammalian cells. The toxic effects of glyphosate or its commercial preparation are essentially associated with the induction of oxidative stress, although increased calcium permeability in the cells and the modulation of intracellular signaling pathways leads to depletion of GSH (reduced glutathione) and modification of the activity of catalase (CAT), glutathione peroxidase (GPx), and superoxide dismutase (SOD) enzymes (Schimpf et al., 2017; Gehin; Guyon; Nicod, 2006).

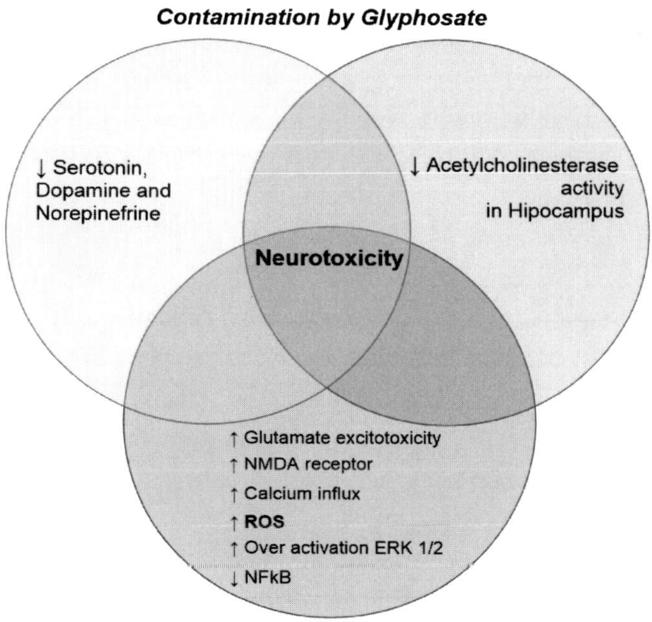

Figure 4. A proposed mechanism of toxicity by glyphosate in the neural system.

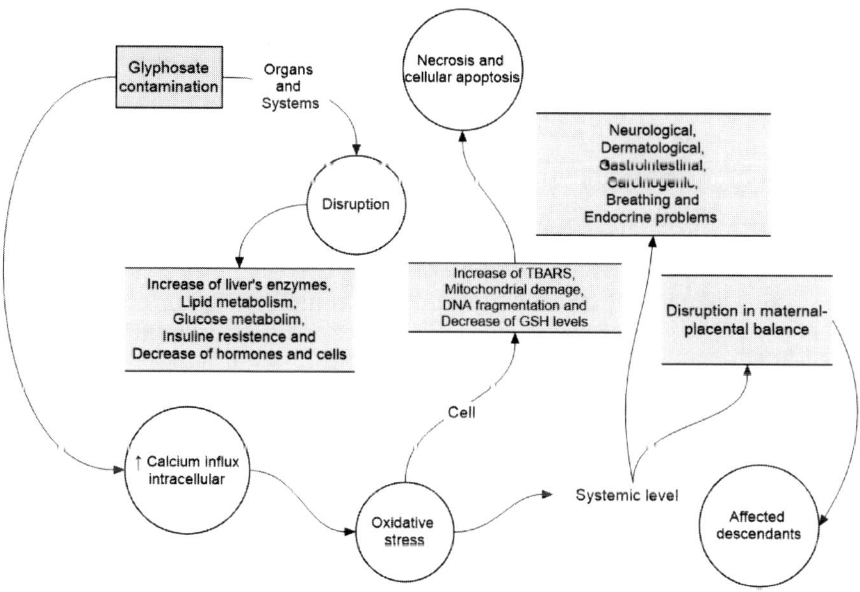

Figure 5. Main effects of contamination by glyphosate.

REFERENCES

A Polyzos, S., Kountouras, J., Deretzi, G., Zavos, C. and S Mantzoros, C. (2012). The emerging role of endocrine disruptors in pathogenesis of insulin resistance: a concept implicating nonalcoholic fatty liver disease. *Current molecular medicine*, 12(1): 68-82.

Aldridge, J. E., Seidler, F. J., Meyer, A., Thillai, I. and Slotkin, T. A. (2003). Serotonergic systems targeted by developmental exposure to chlorpyrifos: effects during different critical periods. *Environmental health perspectives*, 111(14): 1736.

Amaral, A., Lourenço, B., Marques, M. and Ramalho-Santos, J. (2013). Mitochondria functionality and sperm quality. *Reproduction*, 146(5): R163-R174.

Anadon, A., Martinez-Larranaga, M. R., Martinez, M. A., Castellano, V. J., Martínez, M., Martin, M. T. and Bernal, J. L. (2009). Toxicokinetics of glyphosate and its metabolite aminomethyl phosphonic acid in rats. *Toxicology letters*, 190(1): 91-95.

Andrés-León, E., Gómez-López, G. and Pisano, D. G. (2017). Prediction of miRNA–mRNA Interactions Using miRGate. *MicroRNA Detection and Target Identification: Methods and Protocols*, 225-237.

Androutsopoulos, VP, Hernández, AF, Liesivuori, J. and Tsatsakis, AM (2013). Uma visão geral mecanicista dos efeitos associados à saúde de baixos níveis de pesticidas organoclorados e organofosforados. *Toxicologia*, 307: 89-94. [A mechanistic overview of the health effects of low levels of organochlorine and organophosphorus pesticides. *Toxicology*]

Aris, A. and Leblanc, S. (2011). Maternal and fetal exposure to pesticides associated to genetically modified foods in Eastern Townships of Quebec, Canada. *Reproductive Toxicology*, 31(4): 528-533.

Aroonvilairat, S., Kespichayawattana, W., Sornprachum, T., Chaisuriya, P., Siwadune, T. and Ratanabanangkoon, K. (2015). Effect of pesticide exposure on immunological, hematological and biochemical parameters in Thai orchid farmers—a cross-sectional study.

International journal of environmental research and public health, 12(6): 5846-5861.

Barbosa, E. R., Leiros da Costa, M. D., Bacheschi, L. A., Scaff, M. and Leite, C. C. (2001). Parkinsonism after glycine-derivate exposure. *Movement Disorders*, 16(3): 565-568.

Baylis, AD (2000). Por que o glifosato é um herbicida global: pontos fortes, fracos e perspectivas. *Pest Management Science*, 56(4): 299-308. [Why glyphosate is a global herbicide: strengths, weaknesses and prospects]

Beecham, J. E. and Seneff, S. (2016). Is there a link between autism and glyphosate-formulated herbicides? *Journal of Autism*, *3*(1): 1.

Benachour, N., Sipahutar, H., Moslemi, S., Gasnier, C., Travert, C. and Séralini, G. E. (2007). Time-and dose-dependent effects of roundup on human embryonic and placental cells. *Archives of environmental contamination and toxicology*, 53(1): 126-133.

Benedetti, A. L., de Lourdes Vituri, C., Trentin, A. G., Domingues, M. A. C. and Alvarez-Silva, M. (2004). The effects of sub-chronic exposure of Wistar rats to the herbicide Glyphosate-Biocarb®. *Toxicology Letters*, 153(2): 227-232.

Benedetti, D., Nunes, E., Sarmento, M., Porto, C., dos Santos, C. E. I., Dias, J. F. and da Silva, J. (2013). Genetic damage in soybean workers exposed to pesticides: evaluation with the comet and buccal micronucleus cytome assays. *Mutation Research/Genetic Toxicology and Environmental Mutagenesis*, 752(1): 28-33.

Bradberry, S. M., Proudfoot, A. T. and Vale, J. A. (2004). Glyphosate poisoning. *Toxicological reviews*, 23(3):159-167.

Cai, W., Ji, Y., Song, X., Guo, H., Han, L., Zhang, F., ... and Xu, M (2017). Effects of glyphosate exposure on sperm concentration in rodents: A systematic review and meta-analysis. *Environmental toxicology and pharmacology*, 55: 148-155.

Casals-Casas, C., and Desvergne, B. (2011). Endocrine disruptors: from endocrine to metabolic disruption. *Annual review of physiology*, 73: 135-162.

Cattani, D., Cavalli, V. L. D. L. O., Rieg, C. E. H., Domingues, J. T., Dal-Cim, T., Tasca, C. I., Barreto Silva F. R. M. and Zamoner, A. (2014). Mechanisms underlying the neurotoxicity induced by glyphosate-based herbicide in immature rat hippocampus: involvement of glutamate excitotoxicity. *Toxicology*, 320: 34-45.

Cattani, D., Cesconetto, P. A., Tavares, M. K., Parisotto, E. B., De Oliveira, P. A., Rieg, C. E. H., ... and Wilhelm Filho, D. (2017). Developmental exposure to glyphosate-based herbicide and depressive-like behavior in adult offspring: implication of glutamate excitotoxicity and oxidative stress. *Toxicology*, 387: 67-80.

Cattelan, M. D. P. (2017). *Avaliação do perfil bioquímico, hematológico, oxidativo e mutagênico e uso de agrotóxicos por trabalhadores rurais do município de Santiago,* RS. Available from: http://dspace.unipampa.edu.br:8080/jspui/handle/riu/1728. [*Evaluation of biochemical, hematological, oxidative and mutagenic profile and use of pesticides by rural workers in the city of Santiago*]

Cavalli, V. L. D. L. O., Cattani, D., Rieg, C. E. H., Pierozan, P., Zanatta, L., Parisotto, E. B., ... and Zamoner, A. (2013). Roundup disrupts male reproductive functions by triggering calcium-mediated cell death in rat testis and Sertoli cells. *Free Radical Biology and Medicine*, 65: 335-346.

Cecchi, A., Rovedatti, M. G., Sabino, G. and Magnarelli, G. G. (2012). Environmental exposure to organophosphate pesticides: assessment of endocrine disruption and hepatotoxicity in pregnant women. *Ecotoxicology and environmental safety*, 80: 280-287.

Chan, P. and Mahler, J. (1992). NTP technical report on the toxicity studies of Glyphosate (CAS No. 1071-83-6) Administered in Dosed Feed to F344/N Rats and B6C3F1 Mice. *Toxicity report series*, 16:1-D3

Chang, C. B., and Chang, C. C. (2009). Refractory cardiopulmonary failure after glyphosate surfactant intoxication: a case report. *Journal of occupational medicine and toxicology*, 4(1): 2.

Chatterjee, S., Basak, P., Chaklader, M., Das, P., Pereira, J. A., Chaudhuri, S. and Law, S. (2013). Pesticide induced marrow toxicity and effects on marrow cell population and on hematopoietic stroma. *Experimental and Toxicologic Pathology*, 65(3):287-295.

Chen, Y. J., Wu, M. L., Deng, J. F. and Yang, C. C. (2009). The epidemiology of glyphosate-surfactant herbicide poisoning in Taiwan, 1986–2007: a poison center study. *Clinical toxicology*, 47(7): 670-677.

Clair, É., Mesnage, R., Travert, C. and Séralini, G. É. (2012). A glyphosate-based herbicide induces necrosis and apoptosis in mature rat testicular cells in vitro, and testosterone decrease at lower levels. *Toxicology in vitro*, 26(2): 269-279.

Coullery, R. P., Ferrari, M. E. and Rosso, S. B. (2016). Neuronal development and axon growth are altered by glyphosate through a WNT non-canonical signaling pathway. *Neurotoxicology*, 52: 150-161.

Curwin, B. D., Hein, M. J., Sanderson, W. T., Striley, C., Heederik, D., Kromhout, H., ...and Alavanja, M. C. (2007). Pesticide dose estimates for children of Iowa farmers and non-farmers. *Environmental research*, 105(3): 307-315.

Dai, P., Hu, P., Tang, J., Li, Y. and Li, C. (2016). Effect of glyphosate on reproductive organs in male rat. *Acta histochemica*, 118(5): 519-526.

De Souza, J. S., Kizys, M. M. L., da Conceição, R. R., Glebocki, G., Romano, R. M., Ortiga-Carvalho, T. M., ...and Chiamolera, M. I. (2017). Perinatal exposure to glyphosate-based herbicide alters the thyrotrophic axis and causes thyroid hormone homeostasis imbalance in male rats. *Toxicology*, 377: 25-37.

Diamanti-Kandarakis, E., Bourguignon, J. P., Giudice, L. C., Hauser, R., Prins, G. S., Soto, A. M., ...and Gore, A. C. (2009). Endocrine-disrupting chemicals: an Endocrine Society scientific statement. *Endocrine reviews*, 30(4): 293-342.

Duttaroy, A., Zimliki, C. L., Gautam, D., Cui, Y., Mears, D. and Wess, J. (2004). Muscarinic stimulation of pancreatic insulin and glucagon release is abolished in m3 muscarinic acetylcholine receptor–deficient mice. *Diabetes*, 53(7): 1714-1720.

Everett, C. J., & Matheson, E. M. (2010). Biomarkers of pesticide exposure and diabetes in the 1999–2004 National Health and Nutrition Examination Survey. *Environment international*, 36(4): 398-401.

Ford, B., Bateman, L. A., Gutierrez-Palominos, L., Park, R. and Nomura, D. K. (2017). Mapping proteome-wide targets of glyphosate in mice. *Cell chemical biology*, *24*(2): 133-140.

Fritschi, L., McLaughlin, J., Sergi, C. M., Calaf, G. M., Le Curieux, F., Forastiere, F., ... and Martin, M. T. (2015). Carcinogenicity of tetrachlorvinphos, parathion, malathion, diazinon, and glyphosate. *Red*, 114(2).

García-García, C. R., Parrón, T., Requena, M., Alarcón, R., Tsatsakis, A. M. and Hernández, A. F. (2016). Occupational pesticide exposure and adverse health effects at the clinical, hematological and biochemical level. *Life sciences*, 145:274-283.

Gasnier, C., Dumont, C., Benachour, N., Clair, E., Chagnon, M. C. and Séralini, G. E. (2009). Glyphosate-based herbicides are toxic and endocrine disruptors in human cell lines. *Toxicology*, 262(3): 184-191.

Gehin, A., Guyon, C. and Nicod, L. (2006). Glyphosate-induced antioxidant imbalance in HaCaT: The protective effect of Vitamins C and E. *Environmental toxicology and pharmacology*, 22(1): 27-34.

Giesy, J. P., Dobson, S. and Solomon, K. R. (2000). Ecotoxicological risk assessment for Roundup® herbicide. In *Reviews of environmental contamination and toxicology* (pp. 35-120). Springer, New York, NY.

Glusczak, L., dos Santos Miron, D., Moraes, B. S., Simões, R. R., Schetinger, M. R. C., Morsch, V. M., and Loro, V. L. (2007). Acute effects of glyphosate herbicide on metabolic and enzymatic parameters of silver catfish (*Rhamdia quelen*). *Comparative Biochemistry and Physiology Part C: Toxicology & Pharmacology*, 146(4): 519-524.

GM Watch. *Glyphosate herbicide disrupts uterine development* (2016). Available from: http://www.gmwatch.org/en/news/latest-news/17047.

Gomes, M. P., Smedbol, E., Chalifour, A., Hénault-Ethier, L., Labrecque, M., Lepage, L., ... and Juneau, P. (2014). Alteration of plant physiology by glyphosate and its by-product aminomethylphosphonic acid: an overview. *Journal of experimental botany*, 65(17): 4691-4703.

Hadrup, N., Svingen, T., Mandrup, K., Skov, K., Pedersen, M., Frederiksen, H. and Vinggaard, A. M. (2016). Juvenile Male Rats Exposed to a Low-Dose Mixture of Twenty-Seven Environmental

Chemicals Display Adverse Health Effects. *PloS One*, 11(9): e016 2027.

Han, H., Zhu, B., Fu, X., You, S., Wang, B., Li, Z. and Yao, Q. (2015). Overexpression of D-amino acid oxidase from Bradyrhizobium japonicum, enhances resistance to glyphosate in Arabidopsis thaliana. *Plant cell reports*, 34(12): 2043-2051.

Hoffman, R. S., Capel, P. D. and Larson, S. J. (2000). Comparison of pesticides in eight US urban streams. *Environmental Toxicology and Chemistry*, 19(9): 2249-2258.

Hori, Y., Fujisawa, M., Shimada, K. and Hirose, Y. (2003). Determination of the herbicide glyphosate and its metabolite in biological specimens by gas chromatography-mass spectrometry. A case of poisoning by Roundup® herbicide. *Journal of Analytical Toxicology*, 27(3): 162-166.

Husain, K. and Ansari, R. A. (1988). Influence of cholinergic and adrenergic blocking drugs on hyperglycemia and brain glycogenolysis in diazinon-treated animals. *Canadian journal of physiology and pharmacology*, 66(8): 1144-1147.

Indirakshi, J., Sunnesh, A., Aruna, M., Reddy, M. H. K., Kumar, A. C., Chandra, V. S. and Kumar, V. S. (2017). Toxic epidermal necrolysis and acute kidney injury due to glyphosate ingestion. *Indian journal of critical care medicine: peer-reviewed, official publication of Indian Society of Critical Care Medicine*, 21(3):167.

Jasper, R., Locatelli, G. O., Pilati, C., & Locatelli, C. (2012). Evaluation of biochemical, hematological and oxidative parameters in mice exposed to the herbicide glyphosate-Roundup®. *Interdisciplinary toxicology*, 5(3):133-140.

Ji, H., Xu, L., Wang, Z., Fan, X. and Wu, L. (2018). Differential microRNA expression in the prefrontal cortex of mouse offspring induced by glyphosate exposure during pregnancy and lactation. *Experimental and therapeutic medicine*, 15(3): 2457-2467.

Jiang, X., Dias, J. A. and He, X. (2014). Structural biology of glycoprotein hormones and their receptors: insights to signaling. *Molecular and cellular endocrinology*, 382(1): 424-451.

Karami-Mohajeri, S. and Abdollahi, M. (2011). Toxic influence of organophosphate, carbamate, and organochlorine pesticides on cellular metabolism of lipids, proteins, and carbohydrates: a systematic review. *Human & experimental toxicology*, 30(9): 1119-1140.

Kim, Y. H., Hong, J. R., Gil, H. W., Song, H. Y., and Hong, S. Y. (2013). Mixtures of glyphosate and surfactant TN20 accelerate cell death via mitochondrial damage-induced apoptosis and necrosis. *Toxicology in Vitro*, 27(1): 191-197.

Larsen, K., Najle, R., Lifschitz, A. and Virkel, G. (2012). Effects of sub-lethal exposure of rats to the herbicide glyphosate in drinking water: glutathione transferase enzyme activities, levels of reduced glutathione and lipid peroxidation in liver, kidneys and small intestine. *Environmental toxicology and pharmacology*, 34(3):811-818.

Lee, C. H., Shih, C. P., Hsu, K. H., Hung, D. Z. and Lin, C. C. (2008). The early prognostic factors of glyphosate-surfactant intoxication. *The American journal of emergency medicine*, 26(3): 275-281.

Lee, J. W., Choi, Y. J., Park, S., Gil, H. W., Song, H. Y. and Hong, S. Y. (2017). Serum S100 protein could predict altered consciousness in glyphosate or glufosinate poisoning patients. *Clinical Toxicology*, 55(5): 357-359.

Li, C., Tang, J., Hu, P., Li, Y. and Win-Shwe, T. T. (2017). Ion Imbalance Is Involved in the Mechanisms of Liver Oxidative Damage in Rats Exposed to Glyphosate. *Frontiers in physiology*, 8:1083.

Li, M. H., Xu, H. D., Liu, Y., Chen, T., Jiang, L., Fu, Y. H. and Wang, J. S. (2016). Multi-tissue metabolic responses of goldfish (*Carassius auratus*) exposed to glyphosate-based herbicide. *Toxicology Research*, 5(4):1039-1052.

Lopes, P. R. S., Pouey, J. L. F. O., Enke, D. B. S., de Camargo, S. G. O., Correa, G. F., Ribeiro, C. L. G. and Soquetta, M. (2010). Efeitos das aflatoxinas sobre os parâmetros eritrocitários de alevinos de jundiá (Rhamdiaquelen). *Revista da FZVA*, 17(1). [Effects of aflatoxins on the erythrocyte parameters of jundiá fingerlings (Rhamdiaquelen). *FZVA Magazine*]

Malhotra, R. C., Ghia, D. K., Cordato, D. J. and Beran, R. G. (2010). Glyphosate–surfactant herbicide-induced reversible encephalopathy. *Journal of Clinical Neuroscience*, 17(11): 1472-1473.

Mallat, E. and Barceló, D. (1998). Analysis and degradation study of glyphosate and of aminomethylphosphonic acid in natural waters by means of polymeric and ion-exchange solid-phase extraction columns followed by ion chromatography–post-column derivatization with fluorescence detection. *Journal of Chromatography A*, 823(1-2): 129-136.

Martínez, M. A., Ares, I., Rodríguez, J. L., Martínez, M., Martínez-Larrañaga, M. R. and Anadón, A. (2018). Neurotransmitter changes in rat brain regions following glyphosate exposure. *Environmental research*, 161: 212-219.

McQueen, H., Callan, A. C. and Hinwood, A. L. (2012). Estimating maternal and prenatal exposure to glyphosate in the community setting. *International journal of hygiene and environmental health*, 215(6): 570-576.

Mechanick, J. I. (2006). Metabolic mechanisms of stress hyperglycemia. *Journal of parenteral and Enteral Nutrition*, 30(2): 157-163.

Menkes, D. B., Temple, W. A. and Edwards, I. R. (1991). Intentional self-poisoning with glyphosate-containing herbicides. *Human & experimental toxicology*, 10(2): 103-107.

Mesnage, R., Arno, M., Costanzo, M., Malatesta, M., Séralini, G. E. and Antoniou, M. N. (2015). Transcriptome profile analysis reflects rat liver and kidney damage following chronic ultra-low dose Roundup exposure. *Environmental Health*, 14(1):70.

Mesnage, R., Defarge, N., De Vendomois, J. S. and Seralini, G. E. (2015). Potential toxic effects of glyphosate and its commercial formulations below regulatory limits. *Food and Chemical Toxicology*, 84: 133-153.

Mesnage, R., Renney, G., Séralini, G. E., Ward, M. and Antoniou, M. N. (2017). Multiomics reveal non-alcoholic fatty liver disease in rats following chronic exposure to an ultra-low dose of Roundup herbicide. *Scientific reports*, 7: 39328.

Meyer, A., Seidler, F. J., Aldridge, J. E. and Slotkin, T. A. (2005). Developmental exposure to terbutaline alters cell signaling in mature rat brain regions and augments the effects of subsequent neonatal exposure to the organophosphorus insecticide chlorpyrifos. *Toxicology and applied pharmacology*, 203(2): 154-166.

Miller, E., Tadros, S. and Cannon, J. (2015). Perfluorooctanoic Acid-Induced Progesterone Synthesis Inhibition is Associated with Decreased Expression of Steroidogenic Acute Regulatory Protein (StAR) mRNA but Not p450 Side-Chain Cleavage mRNA in hCG-Stimulated mLTC-1 Cells. *The FASEB Journal,* 29 (1 Supplement): 685-15.

Modesto, K. A. and Martinez, C. B. (2010). Effects of Roundup Transorb on fish: hematology, antioxidant defenses and acetylcholinesterase activity. *Chemosphere*, 81(6):781-787.

Moreno, N. C., Sofia, S. H. and Martinez, C. B. (2014). Genotoxic effects of the herbicide Roundup Transorb® and its active ingredient glyphosate on the fish Prochilodus lineatus. *Environmental toxicology and pharmacology*, 37(1): 448-454.

Myers, J. P., Antoniou, M. N., Blumberg, B., Carroll, L., Colborn, T., Everett, L. G. and Vandenberg, L. N. (2016). Concerns over use of glyphosate-based herbicides and risks associated with exposures: a consensus statement. *Environmental Health*, 15(1): 19.

Nardi, J., Moras, P. B., Koeppe, C., Dallegrave, E., Leal, M. B. and Rossato-Grando, L. G. (2017). Prepubertal subchronic exposure to soy milk and glyphosate leads to endocrine disruption. *Food and chemical toxicology*, 100: 247-252.

Negga, R., Stuart, J. A., Machen, M. L., Salva, J., Lizek, A. J., Richardson, S. J., ...and Fitsanakis, V. A. (2012). Exposure to glyphosate-and/or Mn/Zn-ethylene-bis-dithiocarbamate-containing pesticides leads to degeneration of γ-aminobutyric acid and dopamine neurons in *Caenorhabditis elegans*. *Neurotoxicity research*, 21(3): 281-290.

Neiva, T. D. J. C., Moraes, A. C. R., Schwyzer, R., Vituri, C. D. L., Rocha, T. R. F., Fries, D. M. And Benedetti, A. L. (2010). In vitro effect of the herbicide glyphosate on human blood platelet aggregation and coagulation. *Revista Brasileira de Hematologia e Hemoterapia*, 32(4): 291-294. [*Brazilian Journal of Hematology and Hemotherapy*]

Nicolopoulou-Stamati, P., Maipas, S., Kotampasi, C., Stamatis, P. and Hens, L. (2016). Chemical pesticides and human health: the urgent need for a new concept in agriculture. *Frontiers in public health*, 4: 148.

NSW EPA. *What Are Pesticides and How Do They Work?* (2013). Available from: http://www.epa.nsw.gov.au/pesticides/pestwhatrhow.htm.

Olorunsogo, O. O. (1990). Modification of the transport of protons and Ca2+ ions across mitochondrial coupling membrane by N-(phosphonomethyl) glycine. *Toxicology*, 61(2): 205-209.

Olorunsogo, O. O., Bababunmi, E. A. and Bassir, O. (1979). Effect of glyphosate on rat liver mitochondriain vivo. *Bulletin of environmental contamination and toxicology*, 22(1):357-364.

Ortiga-Carvalho, T. M., Chiamolera, M. I., Pazos-Moura, C. C. and Wondisford, F. E. (2016). Hypothalamus-pituitary-thyroid axis. *Comprehensive Physiology* 6(3):1387-428.

Panahi, P., Vosough-Ghanbari, S., Pournourmohammadi, S., Ostad, S. N., Nikfar, S., Minaie, B. and Abdollahi, M. (2006). Stimulatory effects of malathion on the key enzymes activities of insulin secretion in langerhans islets, glutamate dehydrogenase and glucokinase. *Toxicology mechanisms and methods*, 16(4): 161-167.

Porreca, I., D'Angelo, F., De Franceschi, L., Mattè, A., Ceccarelli, M., Iolascon, A. and Scarfò, M. (2016). Pesticide toxicogenomics across scales: In vitro transcriptome predicts mechanisms and outcomes of exposure in vivo. *Scientific Reports*, 6:38131.

Potrebić, O., Jović-Stošić, J., Vučinić, S., Tadić, J. and Radulac, M. (2009). Acute glyphosate-surfactant poisoning with neurological sequels and fatal outcome. *Vojnosanitetski pregled*, 66(9): 758-762.

Pournourmohammadi, S., Farzami, B., Ostad, S. N., Azizi, E. and Abdollahi, M. (2005). Effects of malathion subchronic exposure on rat skeletal muscle glucose metabolism. *Environmental toxicology and Pharmacology*, 19(1): 191-196.

Pournourmohammadi, S., Ostad, S. N., Azizi, E., Ghahremani, M. H., Farzami, B., Minaie, B. and Abdollahi, M. (2007). Induction of insulin resistance by malathion: Evidence for disrupted islets cells metabolism and mitochondrial dysfunction. *Pesticide biochemistry and physiology*, 88(3): 346-352.

Rahimi, R. and Abdollahi, M. (2007). A review on the mechanisms involved in hyperglycemia induced by organophosphorus pesticides. *Pesticide biochemistry and physiology*, 88(2): 115-121.

Ribeiro, D. N., Pan, Z., Duke, S. O., Nandula, V. K., Baldwin, B. S., Shaw, D. R. and Dayan, F. E. (2014). Involvement of facultative apomixis in inheritance of EPSPS gene amplification in glyphosate-resistant Amaranthus palmeri. *Planta*, 239(1): 199-212.

Richard, S., Moslemi, S., Sipahutar, H., Benachour, N. and Seralini, G. E. (2005). Differential effects of glyphosate and roundup on human placental cells and aromatase. *Environmental health perspectives*, 113(6): 716.

Romano, M. A., Romano, R. M., Santos, L. D., Wisniewski, P., Campos, D. A., de Souza, P. B., ... and de Oliveira, C. A. (2012). Glyphosate impairs male offspring reproductive development by disrupting gonadotropin expression. *Archives of toxicology*, 86(4): 663-673.

Romano, R. M., Romano, M. A., Bernardi, M. M., Furtado, P. V. and Oliveira, C. A. D. (2010). Prepubertal exposure to commercial formulation of the herbicide glyphosate alters testosterone levels and testicular morphology. *Archives of Toxicology*, 84(4): 309-317.

Romano, R., Souza, P., Nunes, M. and Romano, M. (2012). Perinatal exposure to a commercial formulation of glyphosate reduces the mRNA expression and increases the protein content of beta TSH in the pituitary of male offspring. *Endocrine Abstracts*, 29: 753.

Samsel, A. and Seneff, S. (2015). Glyphosate, pathways to modern diseases IV: cancer and related pathologies. *J. Biol. Phys. Chem*, 15: 121-159.

Sandrini, J. Z., Rola, R. C., Lopes, F. M., Buffon, H. F., Freitas, M. M., Martins, C. D. M. G. and da Rosa, C. E. (2013). Effects of glyphosate on cholinesterase activity of the mussel *Perna perna* and the fish *Danio rerio* and *Jenynsia multidentata*: *In vitro* studies. *Aquatic toxicology*, 130: 171-173.

Schillace, R. V., Skinner, A. M., Pommier, R. F., O'Neill, S., Muller, P. J., Naik, A. M., ... and Pommier, S. J. (2014). Estrogen receptor, progesterone receptor, interleukin-6 and interleukin-8 are variable in breast cancer and benign stem/progenitor cell populations. *BMC cancer*, 14(1): 733.

Schimpf, M. G., Milesi, M. M., Ingaramo, P. I., Luque, E. H. and Varayoud, J. (2017). Neonatal exposure to a glyphosate-based herbicide alters the development of the rat uterus. *Toxicology*, 376: 2-14.

Schug, T. T., Janesick, A., Blumberg, B. and Heindel, J. J. (2011). Endocrine disrupting chemicals and disease susceptibility. *The Journal of steroid biochemistry and molecular biology*, 127(3-5): 204-215.

Séralini, G. E., Clair, E., Mesnage, R., Gress, S., Defarge, N., Malatesta, M. and De Vendômois, J. S. (2014). Republished study: long term toxicity of a Roundup herbicide and a Roundup-tolerant genetically modified maize. *Environmental Sciences Europe,* 26:1-17.

Shadnia, S., Azizi, E., Hosseini, R., Khoei, S., Fouladdel, S., Pajoumand, A. and Abdollahi, M. (2005). Evaluation of oxidative stress and genotoxicity in organophosphorus insecticide formulators. *Human & experimental toxicology*, 24(9): 439-445.

Solakidi, S., Psarra, A. G., Nikolaropoulos, S. and Sekeris, C. E. (2005). Estrogen receptors α and β (ERα and ERβ) and androgen receptor (AR) in human sperm: localization of ERβ and AR in mitochondria of the midpiece. *Human Reproduction*, 20(12): 3481-3487.

Thongprakaisang, S., Thiantanawat, A., Rangkadilok, N., Suriyo, T. and Satayavivad, J. (2013). Glyphosate induces human breast cancer cells growth via estrogen receptors. *Food and Chemical Toxicology*, 59: 129-136.

Ubessi, L. D., Ubessi, C., Kirchner, R. M., da Rosa Jardim, V. M. and Stumm, E. M. F. (2015). Use of protective equipment by farmers who use pesticides in relation to health problems. *Journal of Nursing UFPE on line*, 9(4): 7230-7238.

Vandenberg, L. N., Colborn, T., Hayes, T. B., Heindel, J. J., Jacobs Jr, D. R., Lee, D. H., ... and Zoeller, R. T. (2012). Hormones and endocrine-disrupting chemicals: low-dose effects and nonmonotonic dose responses. *Endocrine reviews*, 33(3): 378-455.

Varayoud, J., Durando, M., Ramos, J. G., Milesi, M. M., Ingaramo, P. I., Muñoz-de-Toro, M. and Luque, E. H. (2017). Effects of a glyphosate-based herbicide on the uterus of adult ovariectomized rats. *Environmental toxicology*, 32(4): 1191-1201.

Varol, E., Ogut, S. and Gultekin, F. (2014). Effect of pesticide exposure on platelet indices in farm workers. *Toxicology and industrial health*, 30(7):630-634.

Wafa, T., Nadia, K., Amel, N., Ikbal, C., Insaf, T., Asma, K. and Mohamed, H. (2013). Oxidative stress, hematological and biochemical alterations in farmers exposed to pesticides. *Journal of Environmental Science and Health, Part B*, 48(12):1058-1069.

Walsh, L. P., McCormick, C., Martin, C. and Stocco, D. M. (2000). Roundup inhibits steroidogenesis by disrupting steroidogenic acute regulatory (StAR) protein expression. *Environmental Health Perspectives,* 108(8): 769.

Wang, G., Fan, X. N., Tan, Y. Y., Cheng, Q. and Chen, S. D. (2011). Parkinsonism after chronic occupational exposure to glyphosate. *Parkinsonism & related disorders*, 17(6): 486-487.

Wang, Y. and Ho, C. T. (2012). Flavour chemistry of methylglyoxal and glyoxal. *Chemical Society Reviews*, 41(11): 4140-4149.

Zhang, F., Pan, L. P., Ding, E. M., Ge, Q. J., Zhang, Z. H., Xu, J. N. and Zhu, B. L. (2017). Study of the effect of occupational exposure to glyphosate on hepatorenal function. *Zhonghuayu fang yixuezazhi* [*Chinese journal of preventive medicine*], 51(7):615-620.

BIOGRAPHICAL SKETCHES

Vilmair Zancanaro

Affiliation: University of Oeste de Santa Catarina – UNOESC, Videira, Santa Catarina, Brazil. University of Alto Vale do Rio do Peixe – UNIARP, Caçador, SC, Brazil

Education: Graduation in Pharmacy and Master in Science and Biotechnology

Business Address: Rua Paese 180, Bairro Universitário, Videira – Santa Catarina – Brasil. CEP: 89560-000

Research and Professional Experience: Teacher at the University of Alto Vale do Rio do Peixe, teaching subjects such as Basic and Clinical Immunology, Clinical Parasitology, Clinical Cytology, Clinical Hematology and Supervised Internship in Clinical Analysis in undergraduate courses. Has experience in the area of Pharmacy and Clinical Analyzes.

Professional Appointments: I am working in University of Alto Vale do Rio do Peixe

Publications from the Last 3 Years:

Bellaver, E. H., Zancanaro, V., & Barp, E. A. (2017). "Comparative Analysis of Drug Disposal Promoted By An Institution Of Higher Education In Partnership With Public And Private Organs Of The Municipality Of Caçador-Sc". [Análise Comparativa De Descarte Medicamentoso Promovido Por Uma Instituição De Ensino Superior Em Parceira Com Órgãos Públicos E Privados Do Município De Caçador-SC]. *Revista Interdisciplinar de Estudos em Saúde*, *5*(2), 106-116. ["Comparative Analysis of Drug Disposal Promoted By an

Institution of Higher Education in Partnership with Public and Private Organs of the Municipality of Caçador-SC". *Interdisciplinary Journal of Health Studies*.]

da Silva Pereira, F. R., Pedrotti, L. A., & Zancanaro, V. (2017). "Incidence Of Intestinal Parasites In A Municipal School Of Basic Education Situated in the Martello Neighborhood in the City Of Caçador / Sc" [Incidência De Parasitoses Intestinais Em Uma Escola Municipal De Educação Básica Situada No Bairro Martello No Município De Caçador/Sc]. *Revista Interdisciplinar de Estudos em Saúde*, *5*(2), 96-105. ["Incidence of Intestinal Parasites in a Municipal School of Basic Education Situated in the Martello Neighborhood in the City of Caçador /SC." *Interdisciplinary Journal of Health Studies*]

De Mello, M., Zancanaro, V., Bellaver, H. Determination of the iron deficiency and megaloblastic anemic profile in pregnant women attended in the Maternal and Infant Public Service of a city in the Middle West of Santa Catarina. *Revista Brasileira de Análises Clínicas*, v. 48, p. 331-336, 2016. [*Brazilian Journal of Clinical Analysis*]

Franzen, J. M., Foppa, T., & Zancanaro, V. (2016). "The Role Of The Master Pharmaceutical Agent In The Prevention Of Gravity". [O Papel Do Farmacêutico Magistral Na Prevenção De Estrias Gravídicas.] DOI: http://dx.doi.org/10.5892/ruvrd.v14i1.2508. *Revista da Universidade Vale do Rio Verde*, *14*(1), 403-410. ["The Role of the Master Pharmaceutical Agent in the Prevention of Gravity". *Magazine of the University of Vale do Rio Verde*]

Franzen, J. M., Vaz, J. G., Zancanaro, V., & de Bitencourt, R. M. (2016). "Low-dose Whey Protein reduces glucose, triglycerides and controls body weight in Wistar rats". [Baixa dose de Whey Protein reduz glicose, triglicérides e controla o peso corporal em ratos Wistar]. *RBONE-Revista Brasileira de Obesidade, Nutrição e Emagrecimento*, *10*(57), 133-144. ["Low-dose Whey Protein reduces glucose, triglycerides and controls body weight in Wistar rats". *RBONE-Brazilian Journal of Obesity, Nutrition and Weight Loss*]

Zancanaro, V., Bordignon, M., Hüntermann, J., & Bellaver, E. H. (2017). "The role of drugs in controlling viral load and CD4 cells in HIV-infected patients in a city in the Midwest of Santa Catarina". [O papel dos medicamentos no controle da carga viral e de células CD4 em Pacientes Com Hiv de Uma Cidade do Meio-oeste de Santa Catarina]. *Revista Univap*, *23*(43), 34-43.

Emyr Hiago Bellaver

Affiliation: University of Oeste de Santa Catarina – UNOESC, Videira, Santa Catarina, Brazil. University of Alto Vale do Rio do Peixe – UNIARP, Caçador, SC, Brazil

Education: Graduation in Pharmacy and Master in Science and Biotechnology

Business Address: Rua Paese 180, Bairro Universitário, Videira – Santa Catarina – Brasil. CEP: 89560-000

Research and Professional Experience: Biomedical clinical pathologist, scientific advisor and member of the Matrix Support Center (NAM) at Hospital Maicé-Caçador and Teacher at the Alto Vale do Rio do Peixe University where he teaches courses in Pharmacy and Biological Sciences.

Professional Appointments: Working in the University of Alto Vale do Rio do Peixe

Publications from the Last 3 Years:

Bellaver, E. H., Zancanaro, V., & Barp, E. A. (2017). "Comparative Analysis of Drug Disposal Promoted By An Institution Of Higher Education In Partnership With Public And Private Organs Of The

Municipality Of Caçador-Sc". [Análise Comparativa De Descarte Medicamentoso Promovido Por Uma Instituição De Ensino Superior Em Parceira Com Órgãos Públicos E Privados Do Município De Caçador-SC]. *Revista Interdisciplinar de Estudos em Saúde*, *5*(2), 106-116. ["Comparative Analysis of Drug Disposal Promoted By an Institution of Higher Education in Partnership with Public and Private Organs of the Municipality of Caçador-SC". *Interdisciplinary Journal of Health Studies*]

de Mello, M, Zancanaro, V., Bellaver, H. Determination of the iron deficiency and megaloblastic anemic profile in pregnant women attended in the Maternal and Infant Public Service of a city in the Middle West of Santa Catarina. *Brazilian Journal of Clinical Analysis,* v. 48, p. 331-336, 2016.

Zancanaro, V., Bordignon, M., Hüntermann, J., & Bellaver, E. H. (2017). "The Role Of Medicines In Control Of Viral Load And Cd4 Cells In Hiv-Patients in a City Of The Middle-West Of Santa Catarina O Papel Dos Medicamentos". [No Controle Da Carga Viral E De Células Cd4 Em Pacientes Com Hiv De Uma Cidade Do Meio-Oeste De Santa Catarina]. *Revista Univap*, *23*(43), 34-43. ["The Role of Medicines in Control of Viral Load and Cd4 Cells in Hiv-Patients in a City of the Middle-West of Santa Catarina O Papel Dos Medicamentos". *Univap Magazine*]

Claudriana Locatelli

Affiliation: University of Oeste de Santa Catarina – UNOESC, Videira, Santa Catarina, Brazil. University of Alto Vale do Rio do Peixe – UNIARP, Caçador, SC, Brazil

Education: Graduation in Pharmacy and PhD in Pharmaceutical Sciences

Business Address: Rua Paese 180, Bairro Universitário, Videira – Santa Catarina – Brasil. CEP: 89560-000

Research and Professional Experience: I have experience in the field of Pharmacy and Biochemistry, working mainly in the research of antitumor, antioxidant and toxic properties of bioactive molecules. In parallel, it has been working in nanotoxicology. I work as a mentor in the programs of Master of the UNOESC in the campus of Videira - Program of Master in Science & Biotechnology and in the campus of Joaçaba Program of Masters in Biosciences and Health.

Professional Appointments: I am working in University of Oeste de Santa Catarina and University of Alto Vale do Rio do Peixe

Publications from the Last 3 Years:

Cordova, C. A., Locatelli, C., Winter, E., Silva, A. H., Zanetti-Ramos, B. G., Jasper, R., ... & Creczynski-Pasa, T. B. (2017). Solid lipid nanoparticles improve octyl gallate antimetastatic activity and ameliorate its renal and hepatic toxic effects. *Anti-cancer drugs*, *28*(9), 977-988.

de Souza Zanchet, M. Z., Nardi, G. M., de Oliveira Souza Bratti, L., Filippin-Monteiro, F. B., & Locatelli, C. (2017). Lycium barbarum Reduces Abdominal Fat and Improves Lipid Profile and Antioxidant Status in Patients with Metabolic Syndrome. *Oxidative medicine and cellular longevity*, 2017.

Nardi, G. M., Januario, A. G. F., Freire, C. G., Megiolaro, F., Schneider, K., Perazzoli, M. R. A., ... & Niero, R. (2016). Anti-inflammatory activity of berry fruits in mice model of inflammation is based on oxidative stress modulation. *Pharmacognosy research*, *8*(Suppl 1), S42.

Noldin, F. V., Locatelli, C., Cordova, Clarissa Amorin Silva, Noldin, A. T., Vanzin, F., Fae, J. D., Buzzi, F. C., Pilati, C., Cechinel Filho, V., Creczynski-Pasa, T. B. (2015). Citotoxicity of N-phenylmaleimide Derivatives and Inhibition of Melanoma Growth in a Preclinical Mouse Melanoma Model. Research and Reviews*: Journal of Pharmacy and Pharmaceutical Sciences*, v. 3, p. 1-11.

Perazzoli, M. R. A., Perondi, C. K., Baratto, C. M., Winter, E., Creczynski-Pasa, T. B., & Locatelli, C. (2017). Gallic Acid and Dodecyl Gallate Prevents Carbon Tetrachloride-Induced Acute and Chronic Hepatotoxicity by Enhancing Hepatic Antioxidant Status and Increasing p53 Expression. *Biological and Pharmaceutical Bulletin, 40*(4), 425-434.

Silva, A. H., Lima Jr, E., Mansilla, M. V., Zysler, R. D., Pisciotti, M. L. M., Locatelli, C., ... & Siccardi, M. (2017). A physiologically based pharmacokinetic model to predict the superparamagnetic iron oxide nanoparticles (SPIONs) accumulation in vivo. *European Journal of Nanomedicine*, *9*(2), 79-90.

Silva, A. H., Lima, E., Mansilla, M. V., Zysler, R. D., Troiani, H., Pisciotti, M. L. M., ... & Winter, E. (2016). Superparamagnetic iron-oxide nanoparticles mPEG350–and mPEG2000-coated: cell uptake and biocompatibility evaluation. *Nanomedicine: Nanotechnology, Biology and Medicine*, *12*(4), 909-919.

Silva, A. H., Locatelli, C., Filho, U. P., Gomes, B. F., de Carvalho Júnior, R. M., de Gois, J. S., ... & Creczynski-Pasa, T. B. (2017). Visceral fat increase and signals of inflammation in adipose tissue after administration of titanium dioxide nanoparticles in mice. *Toxicology and industrial health*, *33*(2), 147-158.

Silva, A. H., Locatelli, C., Filippin-Monteiro, F. B., Martin, P., Liptrott, N. J., Zanetti-Ramos, B. G., ... & Owen, A. (2016). Toxicity and inflammatory response in Swiss albino mice after intraperitoneal and oral administration of polyurethane nanoparticles. *Toxicology letters*, *246*, 17-27.

Winter, E., Locatelli, C., Pietro, A. D., & B Creczynski-Pasa, T. (2015). Recent trends of chalcones potentialities as antiproliferative and antiresistance agents. *Anti-Cancer Agents in Medicinal Chemistry (Formerly Current Medicinal Chemistry-Anti-Cancer Agents)*, *15*(5), 592-604.

Winter, E., Pizzol, C. D., Locatelli, C., & Crezkynski-Pasa, T. B. (2016). Development and evaluation of lipid nanoparticles for drug delivery: study of toxicity in vitro and in vivo. *Journal of nanoscience and nanotechnology*, *16*(2), 1321-1330.

In: Glyphosate
Editor: Ron Quinn
ISBN: 978-1-53614-002-6

Chapter 5

WATER CONTAMINATION BY HERBICIDE GLYPHOSATE: AN OVERVIEW OF TREATMENT OPTIONS

Mohsen M. Nourouzi[1,*], PhD, Luqman Chuah[2], PhD, Atefeh Chamani[1], PhD, Samaneh M. Soleimani[1] and Nikoo Roghani Esfahani[1]

[1]Department of Environment,
Islamic Azad University Isfahan Branch, Isfahan, Iran
[2]Department of Chemical and Environmental Engineering,
Faculty of Engineering, Malaysia, Selangor, Malaysia

ABSTRACT

Glyphosate is a broad-spectrum, non-selective systemic herbicide used for a great variety of crops control. Glyphosate acid and its salts are moderately toxic compounds in the Environmental Protection Agency (EPA) Toxicity Class II, but the widespread and frequent use of Glyphosate has increased concern about potential adverse effects on

* Corresponding Author Email: mo5227@yahoo.com.

human health. Due to more stringent legislation on wastewater discharge and purity of drinking water, attention is given to the development of an effective remediation technology for the decontamination of Glyphosate. Although Glyphosate residues can efficiently be removed by advanced technologies, the treatments are usually expensive and difficult to maintain. Therefore, different methods and options should be considered while it should be noted that most likely a single method is not sufficient and the key treatment is by integrated methods. This review discusses the old-to-recent papers on environmental risk, analysis and variety of feasible remediation to remove Glyphosate from water and wastewater.

Keywords: herbicide, glyphosate, treatment options

1. Introduction

Glyphosate [N-(Phosphonomethyl) glycine] nonselective is a broad-spectrum, non-selective systemic herbicide used for plants control, including grasses, sedges, broad-leaved weeds, and woody plants. It can be used on non-cropland as well as on a great variety of crops. Glyphosate acid and its salts are moderately toxic compounds in the US Environmental Protection Agency (EPA) Toxicity Class II. It is an organophosphorus compound which possesses a carbon-phosphorus (C-P) bond. This C-P bond is resistant to chemical hydrolysis, thermal decomposition, and photolysis. The half-life of Glyphosate in soil is between 3 to 174 days (Petit and Cabridenc, 1995).

2. Global Production of Glyphosate

Glyphosate herbicide was developed for agricultural use by the Monsanto Company in early 1970s. The company introduced Isopropyl-ammonium, Ammonium and Sodium salts of Glyphosate in the 1970 decade. Zeneca introduced Trimesium (trimethylsulfonium) salt of Glyphosate in 1989. The primary Glyphosate producer is the originating company, namely Monsanto, which has dedicated facilities in the USA,

Belgium, Malaysia, Brazil and Argentina (Baylis, 2000, Jiang and Lucy, 2007). There is a considerable number of companies with the capability of producing up to 1,000 ton technical acid of Glyphosate annually. Glyphosate, the active ingredient in Monsanto's *Roundup*, is the most heavily used chemical weed killer in food and agricultural production. This was as a result of the widespread adoption of genetically engineered crops now grown on more than 175 million acres in the United States (U.S.) and more than 440 million acres around the globe. Other companies with their names and locations are listed in Table 1 (Woodburn, 2000).

Table 1. Companies with capability of producing up to 1,000 ton technical acid of Glyphosate annually

Company	Location
Ancom	Malaysia
Barclay Synthesis	Ireland
IPiCi	Italy
Aragonesas	Spain
Herbex	Portugal
Feinchemie Schwebda	Germany
Pinus	Slovenia
Agan	Israel
Oriental and Korag	South Korea
Excel	India
Icona, Reposo and Quimica Estrella	Argentina
Crystal Chemical	Costa Rica
Agroquimicas	Guatemala
Productos Basicos	Mexico
Kuo Ching Chemical and Hui Kwang Chemical	Taiwan

3. COMMERCIAL GLYPHOSATE

The *Roundup Ultra* label states that the active ingredient is "Glyphosate, N-(phosphonomethyl)glycine, in the form of its isopropylamine salt". The term 'Glyphosate' is the common chemical name, whereas 'N-(phosphonomethyl)glycine' is the chemical name that provides information about the actual chemical structure of the herbicide. Regardless of brand, the active ingredient for all Glyphosate products is exactly the same.

Glyphosate is a type of chemical known as weak acid. Weak acids can donate a hydrogen ion to other compounds. When Glyphosate is formulated into a commercial product, the hydrogen ion on the parent weak acid is replaced with a different salt (ion). The salt itself does not have herbicidal properties, but results in a product that is easier to handle, mixes better with other agricultural chemicals, and/or is more effective than the parent weak acid. All Glyphosate products, except *Touchdown*, contain the isopropylamine salt (IPA) of Glyphosate. The particular salt formulation does not significantly affect Glyphosate performance. The name and structure of Glyphosate parent acid and its formulated salt are shown in Table 2 (Sikorski, 1997).

The number of crops with approved Glyphosate use has significantly increased. Glyphosate was used since its first application in plantations, orchards, vineyards and industrial situations, initially as a result of the widespread adoption of minimum and no-tillage conservation practices (Koger et al., 2005).

Table 2. Chemical name and chemical structure of Glyphosate and its salt

Chemical name	Chemical structure
Glyphosate parent acid	$C_3H_8NO_5P$
Vapour Pressure	Negligible
Solubility	12 gL^{-1} in water and insoluble in organic solvent
Soil Sorption	Strong

4. Glyphosate Pollution

The environment can be contaminated due to direct or indirect herbicide release. Environmental contamination from herbicide residues is due to droplet drift (specially by aerial application), wind-blown spread, transport in surface water runoff, snowmelt, volatility, wash-off from plants, release of herbicide during plant decomposition or groundwater flow. A wide variety of pollutants exist in wastewater that is discharged from the manufacture of Glyphosate. Glyphosate wastewater includes spilling, leaking, washdown from the Glyphosate production unit, washwater from the Glyphosate product recovery and byproduct distillation from the final Glyphosate product isolation step (Monnig et al., 1980).

The extent of contamination by Glyphosate compound depends on the means by which Glyphosate moves off-site and the ambient condition (Goldsborough and Crumpton, 1998, Locke et al., 2008). Laitinen et al. (2006) found that Glyphosate residues in soil are detectable after one year of application. However, leaching of Glyphosate into the groundwater is confined by its strong solid-sorption (Borggaard and Gimsing, 2008). New scientific evidence shows that possible harm to human health could begin at ultra-low levels of glyphosate, e.g., 0.1 parts per billion (ppb). Popular foods tested for glyphosate have glyphosate measurement between 289.47 ppb and at levels as high as 1,125.3 ppb (Mesnage et al., 2015).

5. Environmental Risk

Glyphosate is essentially non-toxic to mammals and birds, but fish and invertebrates are sensitive to the herbicide. It was reported that Glyphosate is lethal to amphibians under natural conditions (Relyea, 2005). Arbuckle et al. (2001) observed that exposure to Glyphosate through farm workers has moderately increased the risk of early abortions. The mortality rate due to Glyphosate poisoning was reported to be between 7.5% and 16.7% (Motojyuku et al., 2008). New research shows that *Roundup* causes liver and kidney damage in rats, as reflected in the functional changes of 4,000 genes at only 0.05 parts per billion (ppb) glyphosate equivalent, indicating damage. Additional studies have found that levels as low as 10 ppb can have toxic effects on the liver of fish (Larsen et al., 2012) and cause a significant damage to the livers and kidneys of rats at 700 ppb (Borggaard and Gimsing, 2008), which is the allowable level of glyphosate found in the U.S. drinking water (Kanissery et al., 2015). Environmental Protection Agency (EPA) identified 1 mgL^{-1} as the safe level of Glyphosate concentration in drinking water. Hazard associated with Glyphosate based on ranges of the relative potencies is classified according to Table 3 (Giesy et al., 2000).

To estimate the toxicity of Glyphosate, the values of LC_{50} corresponding to the Glyphosate concentration are listed in Table 4.

Table 3. Toxicity classification for aquatic and avian species

U.S.EPA toxicity classification	Acute aquatic LC50 (mgL-1)	Avian dietary LC50 (mgkg-1)
Practically nontoxic	>100	>5000
Slightly toxic	>10≤100	>1000<5000
Moderately toxic	>1≤10	>500<1000
Highly toxic	>0.1≤1	>50<500
Very highly toxic	≤0.1	<50

Table 4. LC50 values for fish exposed to components of the herbicide Roundup

Product test	Species	Values of LC50 (mgL-1)
Glyphosate	Aquatic invertebrates	>55
Glyphosate	Fish	> 10
Roundup	Aquatic invertebrates	>5.3
Roundup	Fish	> 2.4

The higher toxicity of *Roundup* as compared to its parent component (Glyphosate) is due to a higher toxicity of surfactant in *Roundup* (Giesy et al., 2000).

6. Fate of Glyphosate

As soon as the herbicide enters a system, it involves various physical, chemical, and microbiological fate processes, such as (Borggaard and Gimsing, 2008, Petit and Cabridenc, 1995):

6.1. Sorption

Sorption is herbicide-solids interaction which controls the dissipation of herbicide in the system. The risk of environmental pollution through pesticides leaching out of soils into water bodies is affected by how strong the compound is sorbed into soil. (Sorption is the process by which one substance becomes attached to another and includes both adsorption and absorption). As compared to other pesticides, glyphosate is recorded to have strong sorption characteristics, reducing the risk of leaching. However, several studies have indicated that in some circumstances there is a risk of the glyphosate leaching into deeper soil layers, where it could end up in ground and surface waters (Hagner et al., 2015).

The level of sorption depends on several soil characteristics, including mineral content and type, pH, soil redox conditions, phosphate content (that can compete with glyphosate for sorption sites) and possibly soil organic matter (Kanissery et al., 2015). Solids can be soil, sediment and organic matter. As compared to other herbicides, Glyphosate has a stronger affinity to positive-charge surface solids, such as mineral soils containing Al^{3+} and Fe^{3+} cations.

6.2. Photolysis

Photolysis of herbicide is the decomposition or reaction with other reagent by the solar radiation. However, Glyphosate is stable to photolysis

6.3. Biotransformation

Biotransformation occurs when the original compound is modified by organisms to reduce its toxicity.

6.4. Biodegradation

Biodegradation is defined as the biologically catalysed reduction of chemicals complex. In the case of organic compounds, biodegradation causes the conversion of much of the C, N, P, S, and other elements in the original compound to inorganic products (e.g., CO_2, H_2O, N_2, N_2O and H_2S). The conversion of organic substrates to inorganic is also known as mineralisation (Bailey, 2002).

Figure 1 shows the major metabolites and bacterial degradation pathway of Glyphosate. Microorganisms can utilise Glyphosate as carbon source to produce aminomethylphosphonic acid (AMPA) or as phosphorus source to produce glycine (Solomon et al., 2007). The first step in Glyphosate degradation is the attack of Glyphosate Nitrogen-Carbon bond, resulting in the production of aminomethylphosphonic acid (AMPA). The degradation of AMPA via the phosphorus bond cleavage yields CH_3NH_3. When Glyphosate is removed, metabolite AMPA has been widely reported as a dead-end metabolite (McAuliffe et al., 1990).

Studies examining the rate of glyphosate degradation showed some variability in results, and the process can depend on a range of factors. There is some evidence for the rate of degradation to be correlated with the population size of bacteria in the soils (Gimsing et al., 2004). Overall, sorption of glyphosate onto soil particles is thought to decrease degradation, but glyphosate that has been sorbed can still be degraded by micro-organisms. Rates will vary with topographical features that affect water availability, soil type, and increase with temperature (Stenrød et al., 2006).

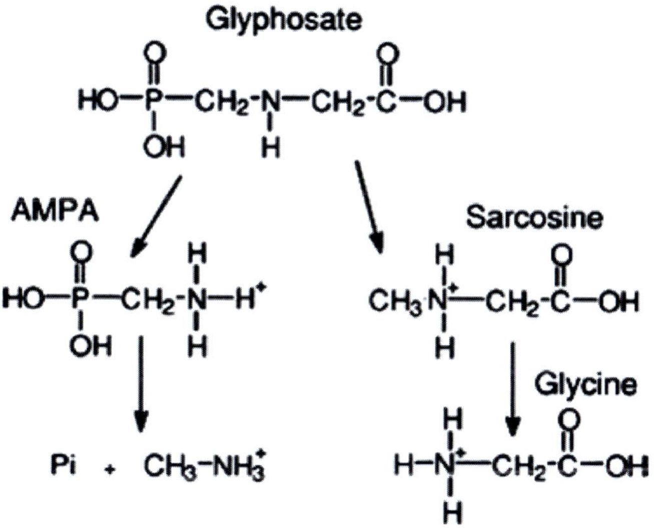

Figure 1. Pathways and metabolites for bio-degradation of Glyphosate.

7. Glyphosate Analysis

Due to environmental contamination risk caused by Glyphosate and the concern about its residual, various methods were developed to detect residues (Stalikas and Konidari, 2001). Glyphosate structure is low in UV absorption, fluorescence, and vaporisation upon heating, the detection of Glyphosate by thin-layer chromatography (TLC), capillary electrophoresis (CE), high performance liquid chromatography (HPLC) or gas chromatography (GC) is difficult (Tseng et al., 2004). Table 5 compares the advantages and disadvantages of each separation method.

To overcome the weaknesses of analytical instrument to detect Glyphosate, a variety of derivatisation methods were developed. Smith (1989) introduced derivatisation of Glufosinate by acetic acid and trimethyl orthoacetate (TMOA) for GC-MS, which was successfully applied for Glyphosate. The reagent N-methyl-N-(tertbutyldimethylsilyl) trifluoroacetamide (MTBSTFA) was used for detecting derivatised Glyphosate by GC- flame photometric detector (FPD) (Moye and Deyrup,

1984) and GC- flame ionisation detector (FID) (Tsunoda, 1993). In HPLC, Glyphosate was derivatised to be detectable by detectors, such as UV visible (Lundgren, 1986), fluorescence (Glass, 1987) and electrogenerated chemiluminescence (Ridlen et al., 1997). Derivatisation can be carried out in either pre-column (prior to injection onto HPLC) or post-column (after separation and prior to detection) modes (McGarvey, 1994).

Table 5. Comparison of current separation technology for Glyphosate detection

Method	Advantage	Disadvantage	Reference
TLC	Ability to use a number of detection methods and reagents in sequence on a single layer.	Low chromatographic separation power. Non-reproducible for quantitative calibration curves.	(Sherma, 2004, Ojanperä, 2000)
CE	Low usage of organic solvent, small sample volume and cost effective as compared to HPLC or GC.	Inadequate limits of detection and a lack of selective detectors.	(Eash and Bushway, 2000, Pico et al., 2003)
HPLC	Useful for small, polar, non-volatile compounds.	Lack of sensitivity of the available HPLC detectors.	(Koch et al., 2010, McGarvey, 1994)
GC	High resolution, rapid separation, low cost and easy linkage with sensitive and selective detectors.	Pre-derivatisation is required.	(Saraji and Bidgoli, 2009, Stalikas and Konidari, 2001)

Recent method in detecting residues is explained here. The sample is extracted by using water and methanol. Extract is further cleaned and concentrated on an ion-exchange column. The extract is then derivatised by using 9-fluorenylmethylchloroformate and filtered prior to LC-MS/MS analysis. Derivatised glyphosate and its metabolite AMPA are injected onto the LC-MS/MS and separated by using a C18 column with a gradient elution of ammonium acetate buffer and acetonitrile as mobile phases.

Ionisation is achieved by using an electrospray ionisation source operating in negative mode. Analytes are detected by using multiple reactions monitoring with isotopically-labelled analytes as internal standards. This method allows detection and quantification down to 2 ng/g (ppb) for food and soil samples with recoveries between 70-90%.

8. Management of Glyphosate Water Pollution

8.1. Bioremediation

Biological treatment is an efficient and cost effective process in remediation of organic matters. Rueppel et al. (1977) studied complete and rapid microbiological degradation of Glyphosate in soil/water shaking flask. They found that up to 50% of each carbon of 1-^{14}C was evolved as $^{14}CO_2$.

Researchers have tried to isolate pure bacterial strains with potential ability for degradation of Glyphosate. These isolates were reported for potential uses in biotechnology industry and biodegradation of polluted soils and water (Moneke et al., 2010). As shown in Figure 1 there are two types of Glyphosate-degrading pathway. The degradation of Glyphosate is affected differently by phosphorus bond. Cleaving of the C–P bond is the first step in the sarcosine pathway, whereas in the AMPA pathway the cleaving of the C–P bond is the second step (Quinn et al., 1988, Borggaard and Gimsing, 2008). Microorganisms known for their ability to degrade Glyphosate in soil and water are listed in Table 6.

Some researchers reported the removal of Glyphosate from aqueous system processed in secondary treatment activated sludge (AS) biosystems. Hallas et al. (1988) suggested that in general, industrial-AS had higher numbers of bacteria than the domestic-AS but fewer pseudomonads and not any yeast. Industrial-AS bacteria were also more competent than domestic-AS bacteria. (Hallas et al., 1992) used a pilot plant to treat waste containing 50 mgL^{-1} of Glyphosate. The up-flow system was configured with the immobilised cell column. The Glyphosate removal of >90% was obtained at a hydraulic residence time (HRT) of 20 min (70 hydraulic turnovers per day). (Castro et al., 2007) also conducted bioremediation of Glyphosate experiments in a batch reactor.

Table 6. Bacterial strains able to degrade herbicide Glyphosate

Species	Source of bacteria	Degradation pathway	Reference
Strain PG2982	Stock culture of Pseudomonas aeruginosa ATCC 9027	Sarcosine	(Moore et al., 1983)
Flavobacterium sp.	Industrial aerobic bio-system (containing Glyphosate wastes)	Sarcosine	(Balthazor and Hallas, 1986)
Arthrohacter sp. GLP-1	Stock culture	Sarcosine	(Pipke et al., 1987)
Pseudomonas sp. strain LBr	Glyphosate process waste stream	Sarcosine	(Jacob et al., 1988)
Pseudomonas spp. and Bacillus megaterium	Soil with long term application of herbicide Glyphosate	Sarcosine	(Quinn et al., 1989)
Agrobacterium radiobacter and Achromobaeter Group V D	Activated sludge	AMPA	(McAuliffe et al., 1990)
Rhizobiaceae family	Stock culture	Sarcosine	(Liu et al., 1991)
Pseudomonas fluorescens		AMPA	(Zboinska et al., 1992)
Xanthomonas maltophilia.	Activated sludge	AMPA	(Carson et al., 1997)
Geobacillus caldoxylosilyticus	Stock culture	Sarcosine	(Obojska et al., 2002)
Stenotrophomonas maltophilia and Providencia alcalifaciens	Soil with long term application of herbicide Glyphosate	AMPA	(Nourouzi et al., 2011)

Their results showed that more than 41% removal of Glyphosate from the batch system was obtained. They also reported that the biodegradation conducted in the bioreactor was more efficient than in the platform shaker. The results also showed that the rate of oxygen had no significant effect on biodegradation process.

8.2. Chemical Treatment

8.2.1. Ferric Chloride Precipitation

Heitkamp et al. (1992) reported the use of ferric chloride for removing Glyphosate from the effluent of Glyphosate manufacturing process. The process was based on the conversion of dissolved Glyphosate into insoluble Glyphosate salt by the addition of ions, such as iron (III), calcium, magnesium or aluminium. The conversion process involves three steps: (i) precipitation in the form of complex salt; (ii) co-precipitation of soluble Glyphosate that is attached onto a growing hydroxide phase; and (iii) adsorption onto the surface complexes of hydroxide. It is reported that iron (III) ions are the most effective due to the formation of least soluble complex salts and consequently are easily removed. The solubility of complex salt is significantly affected by the pH of the solution (McBride and Kung, 1989). Hodgkinson (1999) showed that >94% of the total carbon of Glyphosate solution was removed by adding 7.7 g of a 50% w/w solution of ferric chloride hexahydrate.

8.2.2. Photochemical Oxidation

Electrochemical process is an efficient oxidising condition which is carried out in chloride media to maximise the oxidation of the organic compound at dimensionally stable anode (DSA). The main advantage of the electrochemical treatment is its versatility, easy operation, and lower cost (Gandini et al., 2000, Trasatti, 2000). (Aquino Neto and De Andrade, 2009) investigated the electrochemical degradation of different Glyphosate herbicide formulations on RuO2 and IrO2 DSA. They found that different composition of metallic oxide electrodes containing RuO2 and IrO2 showed similar efficiency concerning herbicide degradation. The results showed that the formation of organochloride had a direct relation with anode composition. They also found that higher NaCl concentrations resulted in a significant increase in Glyphosate mineralisation. But, there was a direct relation between NaCl and concentration and organochloride formation. The commercial herbicide formulations showed higher resistance to oxidation process and also produced a larger quantity of organochloride compounds.

8.2.3. Electro-Fenton Oxidation

Electro-Fenton process is based on the action of oxidant hydroxyl radical [E°(•OH/H2O) = 2.8 V/NHE]. Fenton's reaction between iron catalyst and hydrogen peroxide (i.e., Fenton's reagent) is as shown below:

$$Fe^{2+} + H_2O_2 + H^+ \rightarrow Fe^{3+} + H_2O + {}^{\bullet}OH \quad (1)$$

Balci et al. (2009) reported the use of modified electro-Fenton-like (EF-like) process to degrade aqueous solutions of Glyphosate. Due to the difficulties associated with the formation of stable complexes in the presence of phosphonates, (Balci et al., 2009) proposed the use of Mn^{2+} instead of Fe^{2+}. In this study, they found that Glyphosate was completely destroyed. They also found that the primary byproduct of Glyphosate, AMPA was also completely mineralized. In another study, Navee and Kim (2009a) introduced a sedimentation process by polyaluminum chloride (PAC) and found that more than 60% of Glyphosate was removed from drinking water.

8.2.4. Ozone Oxidation

Ozone is a strong oxidant which is generally produced by an electric discharge method in the presence of air or oxygen (Chiron et al., 2000). Many organic compounds can be easily degraded by ozone. However, phosphonate compound, such as Glyphosate exhibits a complex pH dependence that can be explained by the protonation of the amine group at pH < 9 (David Yao and Haag, 1991, Haag and Yao, 1992). On the other hand, Glyphosate and its main metabolite AMPA, can be formed during the ozonation of complex agent ethylenediaminetetra (methylenephosphonic acid) (EDTMP) (Klinger et al., 1998). Assalin et al. (2010) found that the Glyphosate was degraded more effectively at alkaline pH. Their results showed that after 17 minutes of ozonation (O_3/pH 10), the Glyphosate was totally removed. Similar results were reported by Speth (1993).

8.2.5. Photocatalytic Oxidation

Glyphosate in aqueous system is considerably stable to the long-waved light while UV-light has a strong effect on its photo-degradation (Lund-HØie and Friestad, 1986). (Chen et al., 2007a)$_a$ found that Glyphosate was easily degraded in aqueous solution by using photo catalyst TiO_2 under UV light irradiation. The optimum degradation of the Glyphosate was obtained with TiO_2 concentration of 6.0 $gL^{-1}$and with addition of 0.0196, 0.01, 0.1, 1.0 and 0.5 mM of Fe^{3+}, Cu^{2+}, H_2O_2, $K_2S_2O_8$ or $KBrO_3$, respectively. (Chen et al., 2007b)$_b$ investigated the photodegradation of Glyphosate in the ferrioxalate system under irradiation with a 250W metal halide lamp ($\lambda \geq$ 365 nm). In this study they found that degradation of Glyphosate increased with the decrease of Fe(III)/oxalate ratio at pH of 5. In another study, (Manassero et al., 2010) reported that the combined hydrogen peroxide and UV radiation (H_2O_2/UV). System was an efficient and simple process to remove Glyphosate from water.

8.3. Physical Treatment

8.3.1. Adsorption

The adsorption of pesticides on solid substrates, such as soils, clays, microorganisms or activated carbon has been recognised as one of the effective techniques for the pesticide removal of from water. (Navee and Kim, 2009b) investigated the removal efficiency of Glyphosate by using different processes. In contrast to the sodium hypochlorite treatment, coagulation by polyaluminum chloride and ozonation methods, the amount of Glyphosate adsorbed by granular activated carbon was negligible. In similar studies activated carbon was tested for its ability to remove

Glyphosate from water. Table 7 shows adsorption capacity of the Glyphosate onto different types of activated carbon.

Different adsorbents such as pure iron-humic acid (Fe-HA) complex, MgAl-layered double hydroxides and alum sludge with the average adsorption capacity of 88, 270 and 101 mgL^{-1}, respectively also were used to remove Glyphosate from water (Piccolo et al., 1992, Li et al., 2005, Hu et al., 2011). (Nomikou et al., 2006) described a process where electric field was used to enhance the biosorption of Glyphosate onto seaweed kelp.

Table 7. Adsorption capacity of the Glyphosate onto three different types of activated carbon

Type	Form	Adsorption capacity (mgL^{-1})	Reference
F400	Granule	110	(Speth, 1993)
Commercial	Granule	58.4	(Li-jing et al., 2006)
Waste newspaper activated carbon	Powder	48.4	(Nourouzi et al., 2010)

8.3.2. Membrane

Nanofiltration (NF) is a pressure-driven membrane process that is able to reject organic molecules, such as herbicide Glyphosate while mineral salts with monovalent anions easily pass through the membrane (Xie et al., 2010). (Xie and Xu, 2011) studied four different NF operation modes, including direct nanofiltration, diafiltration, dilute-diafiltration and interval washing-nanofiltration. Results showed that under the applied pressure of 3.0±0.1MPa and during the 4-h direct nanofiltration period, the permeate flux decreased gradually, but the Glyphosate rejection remained more than 90%. Diafiltration and dilute-diafiltration demonstrated better performance than direct nanofiltration. The interval washing-nanofiltration mode was effective in reducing membrane fouling and prolonging the running time for Glyphosate. Dilute-diafiltration was found to be the best operation mode in respect of Glyphosate recovery, salt removal and cost.

8.4. Integrated Treatment

In many studies, researchers have shown that an integrated treatment could synergise the advantages of their treatment performance, while overcoming their respective limitations (Chandrasekara Pillai et al., 2009). Table 8 summarises some combined methods for pesticides wastewater treatment.

Table 8. Integrated treatment of pesticide wastewater

Target compound	Integrated method	Reference
Glyphosate	Reverse osmosis-activated carbon filtration	(Schoonenberg et al., 2010)
Atrazine	Photo-assisted and electrochemical	(Malpass et al., 2010)
Arsenic	Polymeric binding and membrane	(Legault et al., 1993)
Metribuzin	Electrochemical and ultraviolet oxidation	(Yahiaoui et al., 2011)
Triazinone	Anodic-Fenton and Biodegradation	(Scherer et al., 2004)
Phosmet	Electrochemical-biological	(Salles et al., 2010)
Phenyl-urea	Chemical Oxidation-Filtration	(Benitez et al., 2010)
Vydate, Metomur, Couraze, Ditimur-40 and Scala	Photo-Fenton and MBR	(Lopez et al., 2010)
Malathion	Photo-Fenton and nanofiltration	(Zhang and Pagilla, 2010)

8.5. Another Treatment Methods

In literature, a large and growing body of research was directed towards glyphosate treatment. As outlined by Tran et al. (2017), surface and ground waters are directly and indirectly exposed to acute concentrations of glyphosate leading to substantial adverse impacts on living organisms and ecosystems. Currently, a variety of procedures has been suggested for the removal of glyphosate in aquatic environments. For example, Tran et al. (2017) proposed a novel and advanced oxidation process to deal with the challenges related to degradation and mineralisation of glyphosate. Their main idea was to substitute sodium sulphate for chloride, as a common supporting electrolyte, through electro-oxidation of glyphosate. The results showed that sodium sulphate (0.01 mol/L) could substantially increase electrical conductivity of water and in turn provide a cost, time and energy-effective procedure for better degradation of glyphosate in contaminated waters.

Samuel et al. (2017) described a novel method for efficient removal of glyphosate from aqueous media via adsorption onto highly functionalised star-shaped polymeric particles (Amine-functionalised). Ionic interaction between the protonated amines and glyphosate's dissociated carboxylic and phosphoric acid groups lead to effective glyphosate capture even at feed concentrations below 1 ppm. NMR studies conducted with butylamine as a proxy for these polymeric particles confirm that the amine group binds with both glyphosate's carboxylic and phosphoric acid groups when its concentrations are in a 2:1 or higher molar ratio with glyphosate (2017).

Another study showed that single photolysis is an inefficient technology for the removal of the pesticide; however, when coupled with electrolysis the removal yield is significantly improved. The use of a combined process (photo-electrolysis) leads to the generation of higher concentrations of free radicals from the photo-activation of the oxidants electrogenerated. A major finding is that the supporting electrolyte plays a key role on the removal of glyphosate due to the generation of different oxidant species. Such species (peroxocarbonates, peroxosulphates and hypochlorite) also contribute to the depletion of the pesticide (Rubí-Juárez et al., 2016).

Also Carneiro et al. (2015) provided a test in order to remove of glyphosate herbicide from water by using biopolymer membranes with different adsorbents. The results showed that the glyphosate herbicide can be adsorbed by chitosan membranes and the proposed membrane-based methodology was successfully used to treat a water sample contaminated with glyphosate.

In comparison to degradation techniques, some studies attempted to remove (or absorb) glyphosate from water. For instance, Samuel et al. (2017) tested and compared the efficiency of amine-functionalized star polymer (SP-PDMAEMA), PAMAM6 and activated carbon absorbents in capturing glyphosate from 10 ppm solution made by dissolving 0.01 g glyphosate powder in 1 litre deionised water. The results of their experiments revealed the order of SP-PDMAEMA > PAMAM6 > activated carbon. Accordingly they concluded that higher capacity and relatively shorter contact time of SP-PDMAEMA would be due to its size, geometric structure, density and complex behaviour with herbicides. They also recommend to further investigate the adsorbent efficiency of SP-PDMAEMA in capturing different levels of glyphosate and other herbicides.

In soil environment, Wang et al. (2016) employed dielectric barrier discharge (DBD) plasma in treating glyphosate-contaminated soil. In comparison to non-thermal discharge common methods, this study employed a novel and applicable power discharge methodology which was suggested in previous studies. In their experiments, soils were sampled from farmland areas, dried and then spread on a ground electrode. By conducting the DBD plasma extraction under various discharge voltage, glyphosate concentration and its molecular structure were analysed. The study showed that utilising DBD procedure could substantially facilitate glyphosate degradation and would be an alternative and energy effective procedure in the remediation of glyphosate- contaminated soil.

Zheng et al. (2016) increased the adsorption capacity and mineralisation performance of MnOx/Al2O3, as a well-known glyphosate adsorbent. To do so, MnOx/Al2O3 samples were exposed to 1g/L glyphosate solutions, dried and finally regenerated by using microwave (MV), ultraviolet (UV) and the combination of microwave- ultraviolet

(MV- UV) systems. The results of their experiments revealed that by only using UV radiation could not mineralise glyphosate. Moreover, they showed that, in comparison to using microwave reactors, MV- UV could substantially improve the mineralisation performance of MnOx/Al2O3 up to 30 percent. Vidal et al. (2015) carried out a study about the potential effect of hydrogen peroxide and UV irradiation on degradation of glyphosate. Their experiments include 1 micromole hydrogen peroxide at pH 7.0 and 25 °C which were spectrophotometrically analysed at 350 nm wavelength. Comparison of initial and final glyphosate concentrations showed that UV radiated hydrogen peroxide could considerably degrade glyphosate from water. They showed that hydrogen peroxide, a very simple and easy to implement adsorbent, is highly correlated to the amount of glyphosate degradation. Accordingly, with the increase in the amount of UV-radiated hydrogen peroxide, its performance in degrading glyphosate would increase significantly. Herath et al. (2016) investigated the effect of two biochars, including a waste by product of bioenergy plant and a steam activated rice husk derived biochars, termed as DBC and SRBC, respectively on glyphosate physisorption and chemisorption (sorbent doze of 0.5g/L). Accordingly, the optimal pH of the solutions were adjusted with 1 M HNO3 and centrifuged at 3000 rpm for 15 min. They outlined that adsorption performance of biochars in capturing glyphosate is highly dependent to pH and the best performance could be achieved in lower pH values. Furthermore, both biochars were highly effective in adsorbing glyphosate, especially when pH was decreased. Sviridov et al. (2015) comprehensively reviewed the literature published on biodegradation of glyphosate. They discussed that microbial mineralisation, adsorption and transformation is the most rapid and broadly applicable of glyphosate removal. Furthermore, a systematic classification is prepared for glyphosate-degrading microorganisms and the corresponding environments (soil and water). In total, it is concluded that although bacterial glyphosate degradation is slow, environment dependent and mostly has low performance, they could be more useful for safe remediation of natural ecosystems.

CONCLUSION

Glyphosate industry wastewater contains a wide variety of pollutants generated in the manufacturing process. Therefore, the treatment process should take into account the impact and interference of all components present in the wastewater.

Although Glyphosate can be easily degraded to its metabolites, complete bio-degradation has not been reported. On the other hand, Glyphosate degradation may result in production of an even more toxic by-product. Thus, additional treatment is needed to reduce the toxicity of biological treatment effluent.

Glyphosate can be degraded by various means of oxidation such as chemical oxidation, Electro-Fenton oxidation, photo-catalytic oxidation and use of ozone. Similar to biodegradation, Glyphosate oxidation involves the generation of various by-products. Thus, complete mineralisation of Glyphosate or zero total organic carbon (TOC) cannot be achieved via the oxidation process. Although Glyphosate concentration is possibly reduced below the discharge limits, toxicity of the present residues in effluent should be continuously monitored.

In pesticide manufacturing industries, adsorption process is widely used to treat pesticide active ingredients. The use of activated carbon is limited by the structural group of active ingredient, the cost of adsorbent and the presence of the other components. The adsorption process is more suitable for effluent polishing purpose.

Among the treatment processes studied for the removal of Glyphosate from water, chemical precipitation and membrane are the most efficient methods. It is noteworthy that in these processes the elimination of Glyphosate is accompanied by TOC removal and has no by-product residues. However, the main disadvantage of chemical precipitation is the strong pH-dependency.

In Glyphosate wastewater treatment, no single method could be suggested as a complete process while combining two or more treatment systems used in series operation is encouraged.

References

Aquino Neto, S., De Andrade, A. (2009). Electrochemical degradation of glyphosate formulations at DSA® anodes in chloride medium: an AOX formation study. *Journal of applied electrochemistry,* 39 (10): 1863-1870.

Assalin, M., De Moraes, S., Queiroz, S., Ferracini, V. Duran, N. (2010). Studies on degradation of glyphosate by several oxidative chemical processes: Ozonation, photolysis and heterogeneous photocatalysis. *Journal of Environmental Science and Health, Part B,* 45 (1): 89-94.

Arbuckle, T. E., Lin, Z., & Mery, L. S. (2001). An exploratory analysis of the effect of pesticide exposure on the risk of spontaneous abortion in an Ontario farm population. *Environmental health perspectives,* 109(8), 851.

Bailey, R. A. (2002). *Chemistry of the Environment.* Academic Press.

Balci, B., Oturan, M. A., Oturan, N., Sires, I. (2009) Decontamination of Aqueous Glyphosate, (Aminomethyl). Phosphonic Acid, and Glufosinate Solutions by Electro-Fenton-like Process with Mn2+ as the Catalyst. *Journal of Agricultural and Food Chemistry,* 57 (11): 4888-4894.

Balthazor, T. M., Hallas, L. E. (1986). Glyphosate-degrading microorganisms from industrial activated sludge. *Applied and environmental microbiology,* 51 (2): 432.

Baylis, A. D. (2000). Why glyphosate is a global herbicide: strengths, weaknesses and prospects. *Pest Management Science,* 56 (4): 299-308.

Benitez, F. J., Real, F. J., Acero, J. L. (2010). *Elimination of Phenyl-urea Herbicides in Water Matrices by Combined Chemical Oxidation-Filtration Processes.* World Congress on Engineering and Computer Science (WCECS) San Francisco, USA

Borggaard, O. K., Gimsing, A. L. (2008). Fate of glyphosate in soil and the possibility of leaching to ground and surface waters: a review. *Pest Management Science,* 64 (4): 441-456.

Carneiro, R. T., Taketa, T. B., Neto, R. J. G., Oliveira, J. L., Campos, E. V., de Moraes, M. A., da Silva, C. M., Beppu, M. M. Fraceto, L. F.

(2015). Removal of glyphosate herbicide from water using biopolymer membranes. *Journal of environmental management,* 151: 353-360.

Carson, D. B., Hallas, L. E., Heitkamp, M. A. (1997). Biodegradation of N-phosphonomethyliminodiacetic acid by microorganisms from industrial activated sludge. *Canadian journal of microbiology*, 43 (1): 97-101.

Castro, J., Peralba, M., Ayub, M. (2007). Biodegradation of the herbicide glyphosate by filamentous fungi in platform shaker and batch bioreactor. *Journal of Environmental Science and Health, Part B,* 42 (8): 883-886.

Chandrasekara Pillai, K., Kwon, T. O., Moon, I. S. (2009). Degradation of wastewater from terephthalic acid manufacturing process by ozonation catalyzed with Fe2+, H2O2 and UV light: Direct versus indirect ozonation reactions. *Applied Catalysis B: Environmental,* 91 (1-2): 319-328.

Chen, Y., Wu, F., Lin, Y., Deng, N., Bazhin, N., Glebov, E. (2007a). Photodegradation of glyphosate in the ferrioxalate system. *Journal of Hazardous Materials,* 148 (1-2): 360-365.

Chen, Y., Wu, F., Zhang, X., Deng, N., Bazhin, N. Glebov, E. (2007b) Fe (III)-pyruvate and Fe (III)-citrate induced photodegradation of glyphosate in aqueous solutions. *Journal of Coordination Chemistry,* 60 (22): 2431-2439.

Chiron, S., Fernandez-Alba, A., Rodriguez, A., Garcia-Calvo, E. (2000) Pesticide chemical oxidation: state-of-the-art. *Water Research,* 34 (2): 366-377.

David Yao, C., Haag, W. R. (1991). Rate constants for direct reactions of ozone with several drinking water contaminants. *Water Research,* 25 (7): 761-773.

Eash, D. T., Bushway, R. J. (2000). Herbicide and plant growth regulator analysis by capillary electrophoresis. *Journal of Chromatography A,* 880 (1-2): 281-294.

Gandini, D., Mahe, E., Michaud, P., Haenni, W., Perret, A., Comninellis, C. (2000). Oxidation of carboxylic acids at boron-doped diamond electrodes for wastewater treatment. *Journal of applied electro-chemistry,* 30 (12): 1345-1350.

Giesy, J. P., Dobson, S. Solomon, K. R. (2000). Ecotoxicological risk assessment for Roundup® herbicide. *Reviews of Environmental Contamination and Toxicology,* 35-120.

Gimsing, A. L., Borggaard, O. K., Jacobsen, O. S., Aamand, J., Sørensen, J. (2004). Chemical and microbiological soil characteristics controlling glyphosate mineralisation in Danish surface soils. *Applied Soil Ecology,* 27 (3): 233-242.

Glass, R. L. (1987). Adsorption of glyphosate by soils and clay minerals. *Journal of Agricultural and Food Chemistry,* 35 (4): 497-500.

Goldsborough, L. G., Crumpton, W. G. (1998). Distribution and environmental fate of pesticides in prairie wetlands. *Great Plains Research,* 8 (1): 73-95.

Haag, W. R., Yao, C. C. D. (1992). Rate constants for reaction of hydroxyl radicals with several drinking water contaminants. *Environmental science technology,* 26 (5): 1005-1013.

Hallas, L. E., Hahn, E. M., & Korndorfer, C. (1988). Characterization of microbial traits associated with glyphosate biodegradation in industrial activated sludge. *Journal of Industrial Microbiology & Biotechnology,* *3*(6), 377-385.

Hallas, L., Adams, W., Heitkamp, M. (1992). Glyphosate degradation by immobilized bacteria: field studies with industrial wastewater effluent. *Applied and environmental microbiology,* 58 (4): 1215.

Hagner, M., Hallman, S., Jauhiainen, L., Kemppainen, R., Rämö, S., Tiilikkala, K., & Setälä, H. (2015). Birch (Betula spp.) wood biochar is a potential soil amendment to reduce glyphosate leaching in agricultural soils. *Journal of environmental management*, *164*, 46-52.

Heitkamp, M. A., Adams, W. J., Hallas, L. E. (1992). Glyphosate degradation by immobilized bacteria: laboratory studies showing feasibility for glyphosate removal from waste water. *Canadian journal of microbiology,* 38 (9): 921-928.

Herath, I., Mayakaduwa, S., Vithanage, M. (2016). Potential of Different Biochars for Glyphosate Removal in Water; *Implications for Water Safety.*

Hodgkinson, I. (1999) Process for the preparation of N-phosphono-methylglycine. *Google Patents.*

Hu, Y., Zhao, Y., Sorohan, B. (2011). Removal of glyphosate from aqueous environment by adsorption using water industrial residual. *Desalination,* 271: 150-156.

Jacob, G., Garbow, J., Hallas, L., Kimack, N., Kishore, G., Schaefer, J. (1988). Metabolism of glyphosate in *Pseudomonas* sp. strain LBr. *Applied and environmental microbiology,* 54 (12): 2953.

Jiang, J., Lucy, C. A. (2007). Determination of glyphosate using off-line ion exchange preconcentration and capillary electrophoresis-laser induced fluorescence detection. *Talanta,* 72 (1): 113-118.

Kanissery, R. G., Welsh, A., Sims, G. K. (2015). Effect of soil aeration and phosphate addition on the microbial bioavailability of carbon-14-glyphosate. *Journal of environmental quality,* 44 (1): 137-144.

Klinger, J., Lang, M., Sacher, F., Brauch, H. J., Maier, D., Worch, E. (1998). Formation of glyphosate and AMPA during ozonation of waters containing ethylenediaminetetra (methylenephosphonic acid). *Ozone: science engineering,* 20 (2): 99-110.

Koch, D., Johnson, L., Gehrke, C. W., Sandra, P., Van Breemen, R. B (2010). Chromatography—Advances in Environmental and Natural Products, *Chemical Analysis and Synthesis.* John Wiley Sons, Inc.

Koger, C. H., Shaner, D. L., Krutz, L. J., Walker, T. W., Buehring, N., Henry, W. B., Thomas, W. E., Wilcut, J. W. (2005). Rice (Oryza sativa) response to drift rates of glyphosate. *Pest Management Science,* 61 (12): 1161-1167.

Larsen, K., Najle, R., Lifschitz, A., Virkel, G. (2012). Effects of sub-lethal exposure of rats to the herbicide glyphosate in drinking water: glutathione transferase enzyme activities, levels of reduced glutathione and lipid peroxidation in liver, kidneys and small intestine. *Environmental toxicology and pharmacology,* 34 (3): 811-818.

Legault, A. S., Volchek, K., Tremblay, A. Y., Whittaker, H. (1993). Removal of arsenic from groundwater using reagent binding/ membrane separation. *Environmental progress,* 12 (2): 157-159.

Li-jing, S., Qing-hua, L., Yu-zhi, Y., Yun-hai, S. (2006). Static absorption of glyphosate by active carbon. *Chemical Engineer,* 10.

Li, F., Wang, Y., Yang, Q., Evans, D. G., Forano, C., Duan, X. (2005). Study on adsorption of glyphosate (N-phosphonomethyl glycine) pesticide on MgAl-layered double hydroxides in aqueous solution. *Journal of Hazardous Materials,* 125 (1-3): 89-95.

Liu, C. M., McLean, P., Sookdeo, C., Cannon, F. (1991). Degradation of the herbicide glyphosate by members of the family *Rhizobiaceae. Applied and environmental microbiology,* 57 (6): 1799.

Locke, M. A., Zablotowicz, R. M., Reddy, K. N. (2008). Integrating soil conservation practices and glyphosate resistant crops: impacts on soil. *Pest Management Science,* 64 (4): 457-469.

Lopez, J. L. C., Reina, A. C., Gómez, E. O., Martin, M. M. B., Rodriguez, S. M., Pérez, J. A. S. (2010) Integration of Solar Photocatalysis and Membrane Bioreactor for Pesticides Degradation. *Separation Science and Technology,* 45 (11): 1571-1578.

Lund-HØie, K., Friestad, H. O. (1986). Photodegradation of the herbicide glyphosate in water. *Bulletin of environmental contamination and toxicology,* 36 (1): 723-729.

Lundgren, L. N. (1986). A new method for the determination of glyphosate and (aminomethyl) phosphonic acid residues in soils. *Journal of Agricultural and Food Chemistry,* 34 (3): 535-538.

Malpass, G., Miwa, D., Gomes, L., Azevedo, E., Vilela, W., Fukunaga, M., Guimarães, J., Bertazzoli, R., Machado, S., Motheo, A. (2010). Photo-assisted electrochemical degradation of the commercial herbicide atrazine. *Water science and technology: A journal of the International Association on Water Pollution Research,* 62 (12): 2729.

Manassero, A., Passalia, C., Negro, A. C., Cassano, A. E., Zalazar, C. S. (2010). Glyphosate degradation in water employing the H2O2/UVC process. *Water Research,* 44 (13): 3875-3882.

McAuliffe, K. S., Hallas, L. E., Kulpa, C. F. (1990). Glyphosate degradation by Agrobacterium radiobacter isolated from activated sludge. *Journal of Industrial Microbiology and Biotechnology,* 6 (3): 219-221.

McBride, M., Kung, H. (1989). Complexation of glyphosate and related ligands with iron (III). *Soil Science Society of America journal* (USA).

McGarvey, B. D. (1994). Derivatization reactions applicable to pesticide determination by high-performance liquid chromatography* 1. *Journal of Chromatography B: Biomedical Sciences and Applications,* 659 (1-2): 243-257.

Mesnage, R., Arno, M., Costanzo, M., Malatesta, M., Séralini, G.-E., Antoniou, M. N. (2015). Transcriptome profile analysis reflects rat liver and kidney damage following chronic ultra-low dose Roundup exposure. *Environmental Health,* 14 (1): 70.

Moneke, A., Okpala, G., Anyanwu, C. (2010). Biodegradation of glyphosate herbicide in vitro using bacterial isolates from four rice fields. *African Journal of Biotechnology,* 9 (26): 4067-4074.

Moore, J. K., Braymer, H. D., Larson, A. D. (1983). Isolation of a *Pseudomonas* sp. which utilizes the phosphonate herbicide glyphosate. *Applied and environmental microbiology,* 46 (2): 316.

Motojyuku, M., Saito, T., Akieda, K., Otsuka, H., Yamamoto, I., Inokuchi, S. (2008). Determination of glyphosate, glyphosate metabolites, and glufosinate in human serum by gas chromatography-mass spectrometry. *Journal of Chromatography B,* 875 (2): 509-514.

Moye, H. A., Deyrup, C. L. (1984). A simple single-step derivatization method for the gas chromatographic analysis of the herbicide glyphosate and its metabolite. *Journal of Agricultural and Food Chemistry,* 32 (2): 192-195.

Navee, A., Kim, J.-E. (2009a). *Removal of Herbicide Glyphosate in a Drinking Water Treatment System. 한국환경농학회지 제,* 28 (2).

Navee, A., Kim, J. E. (2009b). *Removal of Herbicide Glyphosate in a Drinking Water Treatment System.* 28 (2).

Nomikou, N., Hughes, P., McHale, A. P. (2006). Use of an electric field assisted biosorption process in the removal of hazardous or precious ionic species from wastewater streams. *Journal of Chemical Technology Biotechnology, 81* (9): 1514-1519.

Nourouzi, M. M., Chuah, T., Choong, T. S. Y. (2010). Adsorption of glyphosate onto activated carbon derived from waste newspaper. *Desalination and water treatment, 24* (1-3): 321-326.

Nourouzi, M. M., Chuah, T. G., Choong, S. Y. T. (2011) Glyphosate utilization as the source of carbon: *Introduction of potential source of soil bacteria E-Journal of Chemistry,* 8.

Obojska, A., Ternan, N. G., Lejczak, B., Kafarski, P., McMullan, G. (2002). Organophosphonate utilization by the thermophile Geobacillus caldoxylosilyticus T20. *Applied and environmental microbiology,* 68 (4): 2081.

Ojanperä, I. (2000) C Pesticides. *Handbook of Analytical Separations,* 2: 391-403.

Petit, V., Cabridenc, R. (1995). Review of strategies for modelling the environmental fate of pesticides discharged into riverine systems. *Environment international,* 21 (2): 167-176.

Piccolo, A., Celano, G., Pietramellara, G. (1992). Adsorption of the herbicide glyphosate on a metal-humic acid complex. *Science of the total environment,* 123: 77-82.

Pico, Y., Rodriguez, R., Ma es, J. (2003). Capillary electrophoresis for the determination of pesticide residues. *TrAC Trends in Analytical Chemistry,* 22 (3): 133-151.

Pipke, R., Amrhein, N., Jacob, G. S., Schaefer, J., Kishore, G. M. (1987). Metabolism of glyphosate in an *Arthrobacter* sp. GLP 1. *European Journal of Biochemistry,* 165 (2): 267-273.

Quinn, J. P., Peden, J. M. M., Dick, R. E. (1988). Glyphosate tolerance and utilization by the microflora of soils treated with the herbicide. *Applied microbiology and biotechnology,* 29 (5): 511-516.

Quinn, J. P., Peden, J. M. M., Dick, R. E. (1989). Carbon-phosphorus bond cleavage by Gram-positive and Gram-negative soil bacteria. *Applied microbiology and biotechnology,* 31 (3): 283-287.

Relyea, R. A. (2005). The lethal impact of Roundup on aquatic and terrestrial amphibians. *Ecological Applications,* 15 (4): 1118-1124.

Ridlen, J. S., Klopf, G. J., Nieman, T. A. (1997). Determination of glyphosate and related compounds using HPLC with tris (2, 2'-bipyridyl) ruthenium (II) electrogenerated chemiluminescence detection. *Analytica chimica acta,* 341 (2-3), 195-204.

Rubí-Juárez, H., Cotillas, S., Sáez, C., Cañizares, P., Barrera-Díaz, C., Rodrigo, M. A. (2016). Use of conductive diamond photo-electrochemical oxidation for the removal of pesticide glyphosate. *Separation and Purification Technology*, 167: 127-135.

Rueppel, M. L., Brightwell, B. B., Schaefer, J., & Marvel, J. T. (1977). Metabolism and degradation of glyphosate in soil and water. *Journal of Agricultural and Food Chemistry,* 25(3), 517-528.

Salles, N. A., Fourcade, F., Geneste, F., Floner, D., Amrane, A. (2010). Relevance of an electrochemical process prior to a biological treatment for the removal of an organophosphorous pesticide, phosmet. *Journal of Hazardous Materials,* 181 (1-3): 617-623.

Samuel, L., Wang, R., Dubois, G., Allen, R., Wojtecki, R., La, Y.-H. (2017). Amine-functionalized, multi-arm star polymers: A novel platform for removing glyphosate from aqueous media. *Chemosphere,* 169: 437-442.

Saraji, M., Bidgoli, A. A. H. (2009). Single-drop microextraction with in-microvial derivatization for the determination of haloacetic acids in water sample by gas chromatography-mass spectrometry. *Journal of Chromatography A,* 1216 (7): 1059-1066.

Scherer, E. M., Wang, Q. Q., Hay, A., Lemley, A. (2004). The binary treatment of aqueous metribuzin using anodic Fenton treatment and biodegradation. *Archives of environmental contamination and toxicology,* 47 (2): 154-161.

Schoonenberg, K. F., Rietman, B., Verliefde, A. (2010). Reverse osmosis followed by activated carbon filtration for efficient removal of organic micropollutants from river bank filtrate. *Water science and technology: a journal of the International Association on Water Pollution Research,* 61 (10): 26-03.

Sherma, J. (2004). Planar chromatography. *Analytical chemistry,* 76 (12): 3251-3262.

Sikorski, J. A. (1997). The heterocyclic chemistry associated with the herbicide glyphosate. *Progress in Heterocyclic Chemistry,* 9: 17-42.

Smith, A. E. 1989. Transformation of the herbicide [14C] glufosinate in soils. *Journal of agricultural and food chemistry, 37*(1), pp.267-271.

Solomon, K., Anadón, A., Carrasquilla, G., Cerdeira, A., Marshall, E., Sanin, L. H. (2007). Coca and poppy eradication in Colombia: Environmental and human health assessment of aerially applied glyphosate. *Reviews of Environmental Contamination and Toxicology,* 43-125.

Speth, T. F. (1993). Glyphosate removal from drinking water. *Journal of Environmental Engineering,* 119 (6): 1139-1157.

Stalikas, C. D., Konidari, C. N. (2001). Analytical methods to determine phosphonic and amino acid group-containing pesticides. *Journal of Chromatography A,* 907 (1-2): 1-19.

Stenrød, M., Charnay, M.-P., Benoit, P., Eklo, O. M. (2006). Spatial variability of glyphosate mineralization and soil microbial characteristics in two Norwegian sandy loam soils as affected by surface topographical features. *Soil Biology and Biochemistry,* 38 (5): 962-971.

Sviridov, A., Shushkova, T., Ermakova, I., Ivanova, E., Epiktetov, D., Leont'evskii, A. (2015). Microbial degradation of glyphosate herbicides (review). *Prikladnaia biokhimiia i mikrobiologiia,* 51 (2): 183-190.

Tran, N., Drogui, P., Doan, T. L., Le, T. S., Nguyen, H. C. (2017). Electrochemical degradation and mineralization of glyphosate herbicide. *Environmental technology,* 1-10.

Trasatti, S. (2000) Electrocatalysis: understanding the success of DSA®. *Electrochimica Acta,* 45 (15-16): 2377-2385.

Tsunoda, N. (1993). Simultaneous determination of the herbicides glyphosate, glufosinate and bialaphos and their metabolites by capillary gas chromatography--ion-trap mass spectrometry. *Journal of Chromatography A,* 637 (2): 167-173.

Vidal, E., Negro, A., Cassano, A., Zalazar, C. (2015). Simplified reaction kinetics, models and experiments for glyphosate degradation in water by the UV/H 2 O 2 process. *Photochemical Photobiological Sciences,* 14 (2): 366-377.

Wang, T., Ren, J., Qu, G., Liang, D., Hu, S. (2016). Glyphosate contaminated soil remediation by atmospheric pressure dielectric barrier discharge plasma and its residual toxicity evaluation. *Journal of hazardous materials*, 320: 539-546.

Woodburn, A. T. (2000). Glyphosate: production, pricing and use worldwide. *Pest Management Science,* 56 (4): 309-312.

Xie, M., Liu, Z., Xu, Y. (2010). Removal of glyphosate in neutralization liquor from the glycine-dimethylphosphit process by nanofiltration. *Journal of Hazardous Materials*, 181 (1-3): 975-980.

Xie, M., Xu, Y. (2011). Partial desalination and concentration of glyphosate liquor by nanofiltration. *Journal of Hazardous Materials,* 186, 960-964.

Yahiaoui, O., Aizel, L., Lounici, H., Drouiche, N., Goosen, M. F. A., Pauss, A., Mameri, N. (2011). Evaluating removal of metribuzin pesticide from contaminated groundwater using an electrochemical reactor combined with ultraviolet oxidation. *Desalination,* 270 (1-3): 84-89.

Zboinska, E., Lejczak, B., Kafarski, P. (1992). Organophosphonate utilization by the wild-type strain of Pseudomonas fluorescens. *Applied and environmental microbiology,* 58 (9): 2993.

Zhang, Y., Pagilla, K. (2010). Treatment of Malathion pesticide wastewater with nanofiltration and photo-Fenton oxidation. *Desalination,* 263 (1-3), 36-44.

Zheng, T., Sun, Y., Lin, Y., Wang, N., Wang, P. (2016). Study on preparation of microwave absorbing MnOx/Al 2 O 3 adsorbent and degradation of adsorbed glyphosate in MW–UV system. *Chemical Engineering Journal,* 298, 68-74.

INDEX

A

B

C

D

E

F

G

H

I

K

L

M

N

O

P

Q

R

W

X